AF323724

Archimedes' Stomach... and Other Puzzles You'll Love to Digest

Archimedes' Stomach... and Other Puzzles You'll Love to Digest

Yossi Elran

World Scientific

NEW JERSEY · LONDON · SINGAPORE · BEIJING · SHANGHAI · TAIPEI · CHENNAI

Published by

World Scientific Publishing Co. Pte. Ltd.

5 Toh Tuck Link, Singapore 596224

USA office: 27 Warren Street, Suite 401-402, Hackensack, NJ 07601

UK office: 57 Shelton Street, Covent Garden, London WC2H 9HE

Library of Congress Control Number: 2025010555

British Library Cataloguing-in-Publication Data
A catalogue record for this book is available from the British Library.

ISBN 978-981-12-6859-5 (hardcover)
ISBN 978-981-12-6947-9 (paperback)
ISBN 978-981-12-6860-1 (ebook for institutions)
ISBN 978-981-12-6861-8 (ebook for individuals)

For any available supplementary material, please visit
https://www.worldscientific.com/worldscibooks/10.1142/13207#t=suppl

To my beloved children,
Gilad, Tamar, Racheli, Sharon and Hadas

Preface

Would you like to be able to square numbers in your head? Learn how to make and solve all kinds of quirky mazes? Or perhaps you would like to find out about the origins of Tangram puzzles, decipher some secret codes, dabble in mathematical art and poetry, or solve number puzzles from Ancient Greece and Egypt? If you do, then you've come to the right book. *Archimedes' Stomach ... and Other Puzzles You'll Love to Digest* is a sequel to my first book, *Lewis Carroll's Cats and Rats ... and Other Puzzles with Interesting Tails*, and it delves into a whole new collection of exciting topics in recreational math. It includes a lot of new material, including new insights into some golden oldies, original puzzles that I've crafted, and a glimpse of contemporary works of some of the most creative recreational mathematicians alive today.

The first question you are asked when writing a book is usually, who is this book for? It's a very difficult question to answer. The expected answer might be math and computer science high-school students, undergrads, math teachers, math circle educators — math lovers of all ages. But because I really believe that human beings are naturally very curious and enjoy learning new things and challenging themselves, my answer is: this book is for everyone with a passion for learning. Unfortunately, a lot of the time, peoples' curiosity is quashed by anxiety, an unreasonable fear of not being able to understand. That's why some people shudder even if you just mention the word "math", let alone buy a book about it. But if you've bought this book, this means you've chosen to be "curious", so even if you do get stuck at some point, please don't be intimidated! I've tried to explain things as best as I could, and if you get stuck, you can always email, Facebook, or tweet me if you have any comments or questions.

I wrote this book the same way I like to teach recreational math — associatively. This means that you start off with some kind of bait, and through it, you embark on a journey to discover its history and connection with many surprising topics — not just math — but also art, literature, and magic. The structure of this book is similar to the previous one. Each chapter begins with a puzzle that you are asked to solve, followed by a hint and its solution. Then, we begin our journey into the origins of the puzzle, its historical evolution, the biographies of the main characters involved, and the math topics that are in some aspect connected to it. The next section of the chapter takes a look at how the puzzle can be generalised. After that, there is a short recap of the chapter with some of the main concepts that were discussed, and then a section with some extra problems and challenges and their solutions. The chapters end with a list of references.

This book would not have been possible without the help of many people. First and foremost, World Scientific Publishing, especially Rochelle Kronzek, without whose support, professionalism, and encouragement I would never have finished this book; my editor, Gregory Lee, and Lai Fun Kwong, who was always very accommodating and helpful with all my requests. I would like to thank all those who contributed to this book: Margaret Kepner and Martin Krzywinski, who generously let me use their artwork in the book; Robin Chapman, Madhur Anand, and JoAnne Growney, for their math poetry; Elizabeth Carpenter for her mazes; David Goodman, Ilan Garibi, and World Scientific for the use of Tangram puzzles from *The*

Tangram Book; Kate Jones and Joe Marasco from Kadon Enterprises for sharing their images and research on the Ostomachion, and, Gilroy Song for the Pickagram images. Your invaluable contributions have added so much to the book and allowed me to include more of "the latest" recreational math developments. Thank you all!

Quite a few people reviewed the book or parts of it, and I would like to thank you all for your priceless advice: David Goodman, and Zvi Paltiel, and also some of my close family members — my wife, Michal and my father, Prof. Simon Godfrey.

The Davidson Institute of Science Education at the Weizmann Institute of Science has been my professional home and family for a long time. In the last few years, I have also been lucky enough to find a second professional home as a lecturer at the Education Department of the Western Galilee College in Akko, Israel. At the Davidson Institute, I teach recreational math to kids, while at the Western Galilee College, I teach it to students who I hope to infect with my ongoing enthusiasm, passion, and love for recreational math. I hope they and you might join me and others on our drive to incorporate recreational math into the formal education system. I am very grateful to the Weizmann Institute, the Davidson Institute, and the Western Galilee College for their ongoing support and help.

Lastly, my heartfelt thanks go to my wife, companion, and soul-mate, Michal, whose unwavering inspiration, support, and encouragement have always been a great source of comfort to me. I dedicated my previous book to her, and this book I dedicate to our beloved children, Gilad, Tamar, Racheli, Sharon, and Hadas. I love you all so much! Without you, none of this could have ever, ever, happened.

Eslyv jzf qzc mfjtyr estd mzzv. That's in code, and if you want to learn how to decipher it, go ahead and start reading Chapter 1...

Yossi Elran

March 2025

Contents

Chapter 1

Transposition Ciphers

Introduction

Secret codes are fascinating! They demonstrate the strong connection between language and math, a connection that we will explore later on in the book. Secret codes are used whenever you want to transform some form of intelligible information, a word, message, picture, or whatever, into something incomprehensible to people other than those you choose.

In this chapter, we restrict ourselves to information in the form of text, which, when readable to anyone, we will refer to as *plaintext*. When plaintext is encrypted, it becomes *ciphertext*. The transformation rules are the *key*.

Before we delve any further, we begin with our first puzzle — a secret message you must try and decipher without knowing the key. Cryptanalysts do this process, which is called *breaking the code*.

The Puzzle

Figure 1.1 shows an empty 4 × 24 grid. A nursery rhyme has been encrypted by writing its letters in the grid, ignoring the spaces between the words, and then extracting each of the 4-letter columns in random order. Here is the result (the smiley face is intentional):

JTFA CFLD L☺NR EORU HRSG PTHI LKWE LIDE LANM AOEN NFOM DHWL JANC EJCA WLBT ATDI NCOL TWKB KELJ IPAA HARF ICOT TEIN UAEL

Reconstruct the grid by placing the columns in the grid in the right order so that the nursery rhyme can be read horizontally throughout the grid. By doing this, you are effectively deciphering the encrypted rhyme.

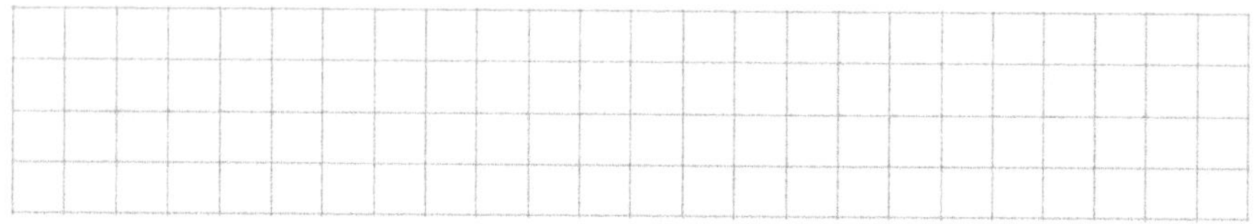

Fig. 1.1   An empty 4 × 24 grid

Where to Start?

To help you along, I've put a smiley face at the end of the second line so you already know the four letters of the last column. You also know that the letter before the smiley face in the 23rd column is the last letter of a word and that the letter in the third row of the first column is the beginning of a new word. Another hint I'm willing to give away is that the first four letters (*JTFA*) are the letters in the first column. Now, try and think of a nursery rhyme that begins with the letter *J* and try to assemble its words in the columns. You might also want to try the following strategy. Try to find columns where two adjacent letters in each row are likely to go together. For example, it would be highly unlikely for a *Q* to come before a *T* in an English word, but not so surprising if it came before a *U*. Be careful, though. If we do not know where a word ends and another begins, we can find some unusual pairs; for example, the pair *FW* might be the end of the word *OF* and the beginning of the word *WATER*.

Solving the Puzzle

Figure 1.2 shows the full solution to the puzzle, and here is the "key" — the order in which the columns were extracted from the grid: 1, 3, 24, 13, 19, 17, 23, 11, 10, 2, 14, 7, 8, 20, 12, 5, 6, 15, 4, 9, 21, 22, 18, 16. Now you can see why I had to add the smiley face — so that the text would completely fill the grid.

1	2	3	4	5	6	7	8	9	10	11	12	13	14	15	16	17	18	19	20	21	22	23	24
J	A	C	K	A	N	D	J	I	L	L	W	E	N	T	U	P	T	H	E	H	I	L	L
T	O	F	E	T	C	H	A	P	A	I	L	O	F	W	A	T	E	R	J	A	C	K	☺
F	E	L	L	D	O	W	N	A	N	D	B	R	O	K	E	H	I	S	C	R	O	W	N
A	N	D	J	I	L	L	C	A	M	E	T	U	M	B	L	I	N	G	A	F	T	E	R

Fig. 1.2   The solution to the puzzle

The History of the Puzzle and its Inventor

Secret codes have a long history. Even the name of ancient Chaldea, mentioned in the Bible, is written in code. Jeremiah, the prophet who witnessed the Babylonian conquest of Jerusalem and the destruction of the first Temple, prophesied the downfall of Nebuchadnezzar II, who was of Chaldean ethnicity. In the first verse of chapter 41, he proclaims: "Thus said the Lord: See, I am rousing a destructive wind against Babylon and the inhabitants of Leb-Kamai." There is one problem with this verse. There is no place known as "Leb-Kamai"! This is the only appearance of the word in the Bible, but surely it is important if its destruction is foreseen along with the Babylonian empire!

The answer is given later on in the chapter, where the word for Babylon is mentioned twice in the same verse — first in code and then correctly. In the first half of the sentence, Babylon is encrypted into a three-letter word ששך, Sheshach. In the second half, it is written correctly as בבל, pronounced "Babel". The whole verse reads:

"How has Sheshach been captured, The praise of the whole earth been taken! How has Babylon become a horror to the nations!"

The cipher method used here is known as "at-bash". To encrypt a word or a message using "at-bash", the Hebrew alphabet is written down from the first letter, aleph, א, to the last letter, tav, ת. Then, in a line underneath, the alphabet is written *back to front*, starting at tav, ת, and ending with aleph, א. These two lines are known as the *key*, the rules for encrypting or deciphering a secret message. The key tells you how to replace a plaintext letter with a cipher letter and vice versa.

In the case of the at-bash cipher, each letter in the top line is replaced by its counterpart in the bottom line. When this is done, the letter bet, ב, is replaced with the letter shin, ש, and the letter lamed, ל, with the letter ך, kaf. Hence, בבל, Babel, becomes, ששך, Sheshach. Back to the Chaldean Nebuchadnezzar II. Using "at-bash", לב-קמי, Leb-Kamai, becomes כשדים, Casdim, the Hebrew pronunciation of Chaldea. No one quite knows why Jeremiah was careful not to mention Chaldea in the first verse, especially since it is explicitly written later in the chapter — four times! Perhaps it was just "prophetic license"!

"At-bash" is an example of what is known as a substitution cipher, and the example above from the bible is arguably the first known use. A substitution cipher is a type of code that replaces letters with other letters, numbers, or symbols. In ancient Rome, Julius Caesar used a substitution cipher to send secret messages to his generals. The Caesar cipher was similar to "at-bash", except that each plaintext letter was encrypted with the letter that comes 3 places after it in the alphabet. If we write the cipher alphabet underneath the plaintext alphabet, it will look like this (note that the alphabet is "wrapped around" so that x, y, and z are encrypted with a, b, and c):

```
A B C D E F G H I J K L M N O P Q R S T U V W X Y Z
D E F G H I J K L M N O P Q R S T U V W X Y Z A B C
```

The word "Hello" in this Caesar cipher is: "Khoor". The sentence: "There are a lot of people on the train" becomes "Wkhuh duh d orw ri shrsoh rq wkh wudlq". Note that the cipher letters are all in the correct order, so the sentence "looks" the same, but the letters have all changed. Extrapolating on this idea, we can generalise and say that letters that appear a lot in the English language, t, h, and e, will appear less in an enciphered text but their counterparts, w, k, and h in the Caesar code will appear more, with the same frequency as the original t, h, and e. Codebreakers exploit this statistical idea to break simple substitution codes. So, once they know that a secret message is a simple substitution cipher, they can break the cipher using statistics even if they don't know the exact key. Since the letter E is the most abundant in English plaintext (assuming that the text is long enough), the letter that appears with the highest frequency in a substitution ciphertext is probably an E. You can see for yourself that even in the Caesar cipher sentence "Wkhuh duh d orw ri shrsoh rq wkh wudlq", the most frequent letter, h, is indeed the code for the letter e.

Statistics and guesswork based on our knowledge of how language works are usually enough to break simple substitution ciphers if the message is long enough. Nowadays, computers use more advanced algorithms to do this, so there is no need for human intervention, at least for simple codes. There are some great online tools for encoding and deciphering substitution messages and I've put links to some in the bibliography at the end of this chapter, but you can even just paste ciphertext into some generative artificial intelligence app and it might do the job itself. I was amazed that ChatGPT did this for the cipher above.

It's important to emphasise that not all substitution of letters with other letters, numbers, or symbols are ciphers. A counter-example is hieroglyphs, a form of writing used in Ancient Egypt that used symbols or pictures to represent words or sounds. These were used not to conceal information but to convey it to people who knew the system. The fact that the system nowadays might seem cumbersome doesn't mean it's a cipher. It's a form of language used by a group of people who learned it, in this case, Egyptian priests. This is similar to the use of Braille for those who are visibly impaired.

Substitution ciphers are well-known, even to kids, and are great fun. Many different substitution ciphers have been devised throughout history, some of them quite complex. However, we now turn to our puzzle and its history, which uses a different kind of cipher system altogether, a transposition cipher.

A transposition cipher is a type of encryption that applies a mathematical function, called a transposition, to the plaintext to produce the ciphertext without changing the actual letters — only their position within the text. One of the main differences between transposition ciphers and substitution ciphers is that substitution ciphers change the letter frequencies and preserve the structure of the plaintext, while transposition ciphers preserve the letter frequencies but change the structure of the plaintext. Substitution ciphers are generally easier to break because the same letter is always replaced with the same letter or symbol. In contrast, in a transposition cipher, the plaintext is transformed to make it harder for an attacker to reverse the process. Transposition ciphers are more ingenious than substitution ciphers because all the letters of the secret message are open to view. They're all there!

The transposition cipher puzzle at the beginning of this chapter is called a *columnar transposition cipher* because the letters of the text are arranged in columns. Although columnar transposition ciphers have been used since ancient times, their exact date of invention remains unknown. We are familiar, though, with a very crude transposition cipher known as the scytale, invented by the Greeks and used by them and the Spartans in the many wars that they held. First, a leather strap was wound around a cylinder. The plaintext message was then written line by line on the leather, as shown in Fig. 1.3. Then, the strap was unwound. The unwound leather strap was the encoded message because it had the vertical placement of the letters on it. This would have been illegible to someone who didn't own an identical cylinder to wrap the strap around. Luckily, the Greeks and Spartans made sure that the recipient owned a cylinder identical to the one used to encode the message so that when he wrapped the strap on *his* cylinder, he could read off the message horizontally.

Fig. 1.3 Image of a scytale with the Latin plaintext Keiser Augustin hars viktet

The scytale is analogous to the columnar transposition cipher, where the number of columns is the number of times the strap is wrapped around the cylinder, the plaintext message is written along the rows, and the cipher is read out down the columns in order.

Figure 1.4 shows the sentence "The key is inside Mary Smiths letterbox" as a columnar transposition cipher and a scytale.

T	H	E	K	E	Y	I	S	I	N	S
I	D	E	M	A	R	Y	S	M	I	T
H	S	L	E	T	T	E	R	B	O	X

Fig. 1.4 A scytale as a columnar transposition cipher

Since the Middle Ages, modifications of columnar transposition ciphers began to appear. One of them, the rail fence transposition cipher, was used by both sides in the American Civil War. The plaintext of a message encrypted using the rail fence cipher is written in zigzag, resembling a rail fence. The ciphertext is read off horizontally. The key is the number of lines the fence is made out of. Figure 1.5 shows an example of a 3-lined rail fence cipher. The plaintext message is: "The raid begins at midnight", the ciphertext is: "TAESIGHRIBGNAMDIHEDITNT", and the key is 3.

T			A		E		S			I		G										
	H	R	I	B	G	N	A	M	D	I	H											
	E		D		I		T		N		T											

Fig. 1.5   A rail fence cipher

At some point, other variations on the columnar transposition cipher were introduced, usually to make the cipher more difficult to crack. One variation was to add blank cells (i.e. cells with no letters) in each column as shown in Fig. 1.6. The plaintext for this cipher is: "The enemy is preparing to attack", and the ciphertext, using the key: 1, 6, 2, 4, 3, 5, is: "TMIACEIETNATTESPGEPROAHYRNK". Arthur Owens, a German double-cross secret agent nicknamed "Snow", working for England during the Second World War, used this transposition cipher variation.

1	6	2	4	3	5
T	H	E	E	N	E
M	Y	I	S		P
	R	E	P	A	R
I	N		G	T	O
A		T		T	A
C	K				

Fig. 1.6   A columnar transposition cipher with windows

Another variation is the route cipher where the ciphertext is extracted according to a pre-determined route within the table, as shown in Fig. 1.7. The plaintext for this cipher is: "Meet me at the restaurant at seven o'clock." The ciphertext is extracted anti-clockwise along a spiral route from the top-right corner to the centre to give: "MTEEMEEATVLOCKXCENTHTTARUAENOSASERT". Since the message is one letter short of the number of cells in the box, an extra letter X is added in the last cell.

Fig. 1.7   A route cipher

Yet another variation was invented by a French general, Émile Victor Théodore Myszkowski, who published a book on cryptography, *Cryptographie indéchiffrable*, in 1902. Myszkowski claimed his cipher was unbreakable, but it is almost as easily broken as the classic columnar transposition cipher. Myszkowski's transposition starts out by generating a key from a given word where at least one of its letters is repeated, for example, the word PAPER. Each letter is turned into a number according to alphabetical order, so PAPER would become: 3, 1, 3, 2, 4. The fact that at least two numbers are the same is what makes it different from the regular columnar transposition cipher. When a message is encoded, the plaintext is written horizontally across the columns (along the rows). The ciphertext is taken vertically from the columns according to the order of the numbers in the key unless a number in the key is repeated, like the number 3 in our example. The letters in each row of these kinds of columns are first merged from left to right before being extracted vertically. So, if, as is the case in our example, there are two columns with the same key, the two columns are merged horizontally, giving pairs of letters that are then extracted vertically. Figure 1.8 shows an example of Myszkowski's transposition of the message: "There is no food left for the troops", which, when extracted, becomes the ciphertext: "HSOTHOROLOTSTEINODFFTEOPEFERRX".

3	1	3	2	4
T	H	E	R	E
I	S	N	O	F
O	O	D	L	E
F	T	F	O	R
T	H	E	T	R
O	O	P	S	X

Fig. 1.8   Myszkowski's transposition

Arguably, the first transposition cipher variation that was one of the more difficult to break was the double columnar transposition that was used by the different parties during both world wars. After encoding the plaintext using one table and key, the cipher is extracted and used as the input for a second table with another key, so the plaintext is doubly encrypted.

My favourite transposition cipher, without a doubt, is the Grille method. The Grille goes back to the 16th century. It is said to have been invented by the Italian mathematician Gerolamo Cardano. Cardano was a jack-of-all-trades and wrote over 200 scientific works on nearly any subject you can imagine — math, physics, medicine, and biology — to name but a few. He was born in 1501 in Pavia,

Italy, just south of Milan, to Chiara Micheri and Fazio Cardano, a gifted lawyer and a friend of Leonardo da Vinci. He had a very tough time throughout his adolescence and adulthood. He was frequently ill, had personality problems, and was constantly harassed for being Micheri and Cardano's illegitimate son. He was also always short of money — and a chronic gambler. He studied science and philosophy at the University of Pavia and medicine at the University of Padua, but due to his personality, he was never accepted to Milan's College of Physicians, despite his undeniable intelligence. Throughout his life, he held several positions, including teaching math and practicing medicine in Milan and being a professor of medicine at the University of Bologna. Unfortunately, wherever he went, he usually ended up at odds with the local scientists, who were envious of his successes, and the local authorities due to his eccentric behaviour. One of the reasons he moved to Bologna in the first place was an accusation of improper sexual behaviour on his part. He even got in trouble with the Inquisition after publishing a book on astrology, which they considered heretic, upon which he was arrested and spent several months in prison. Cardano married and had three children, but his wife died young; one of his children was addicted to gambling so much that he was disinherited, and another poisoned his wife after discovering that their three children were not his — and was duly executed. Gerolamo Cardano's last years were spent as a medical doctor in Rome, where he became a member of the Royal College of Physicians and lived off a generous stipend granted by the Pope. You might be surprised that Gerolamo Cardano's portrait (Fig. 1.9) is displayed at the School of Mathematics and Statistics at the University of St. Andrews in Scotland. There is a good reason for this. Cardano was summoned to St. Andrews in 1553 to treat the archbishop, John Hamilton, who suffered from a disease that was thought to be incurable. Cardano claims in one of his books that he was paid 1,400 gold crowns for his successful treatment.

Fig. 1.9   Portrait of Gerolamo Cardano

Despite his misfortunes, Cardano's scientific accomplishments were many. His inventions include the combination lock, the gimbal, and the drive shaft, appropriately known as Cardan's shaft. In mathematics, he is known for systematically using negative numbers, his anticipation of imaginary numbers, his works on algebra, and his studies on the hypocycloid — the pattern you get when you trace a fixed point on a small circle rolling around the inside a larger circle. Perhaps his largest contribution to math is probability theory, where he was one of the first mathematicians to lay down the foundations of this important field. This should not come as a surprise to us since he used probability to help his gambling! Cardano's grille is an ingenious way to conceal a secret message. Suppose you want to encode the message: "Meet me at four." First, you write the letters in a table far larger than the number of letters in the message (Fig. 1.10, image on the left). Then, fill in all the other cells in the tables with random letters (Fig. 1.10, second image from the left). Take a piece of paper or cardboard the size of the table and cut out windows that correspond to the places where the letters of

M		E	E	
	T	M		E
A		T		
	F	O		
		U		R

M	Q	E	E	R
I	T	M	A	E
A	S	T	O	N
Y	F	O	L	L
O	W	U	N	R

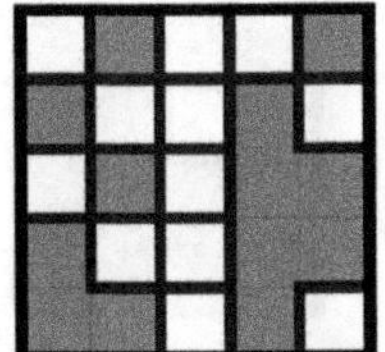

Fig. 1.10   The four stages of the grille encoding of the message: "Meet me at four"

the original message are (Fig. 1.10, third image from the left). When the cardboard grille is placed over the table with the message, the message appears in the windows (Fig. 1.10, image on the right). The table with the secret message can be safely sent to the message's recipient, the only other person with an identical grille, without which the message cannot be read.

The French cardinal Armand Jean du Plessis, Duke of Richelieu, Foreign Minister of France, headmaster of the Sorbonne College, and, incidentally, the inventor of the table knife, used the grille method for his many diplomatic missions. Spies and military officers used the system and its extensions throughout the 17th to 20th centuries. One of the most popular versions of the grille was the rotating grille.

Contrary to the original grille, which wasted a lot of space because of the use of dud letters, the rotating grille used all the cells in the table for plaintext letters. The idea was that the table should be square, and the number of windows in the cardboard grille should be exactly a quarter of the number of cells in the table. The windows were designed so that all the cells in the table were uncovered once the grille was rotated four times. Figure 1.11 shows an example of encoding the message "The tanks are ready" using a rotating grille. We begin with a table of 16 cells to accommodate the 16-letter message. The grille must have 4 windows placed so that, when rotated, each of the 16 cells is exposed once, 4 at each rotation. To write the message, first design the grille and place it above the empty table (Fig. 1.11, first two figures on the top left). Then, write the first 4 letters, T, H, E, and T, in the table through the

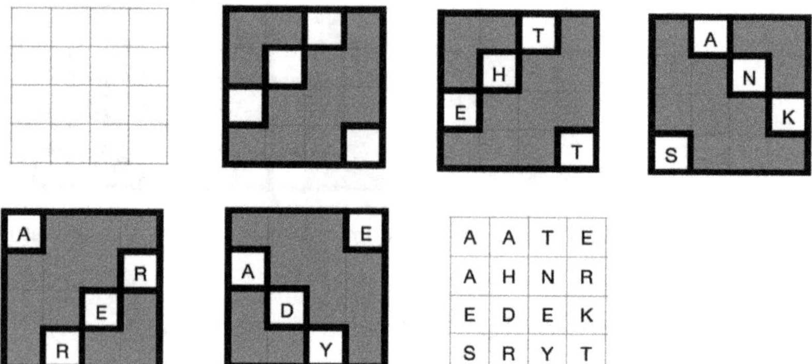

Fig. 1.11   Encoding a message using a rotating grille

four windows (Fig. 1.11, third figure on the top left). Rotate the grille 90° clockwise. Four new blank cells will appear in the windows. Write the next 4 letters, A, N, K, and S, in these cells (Fig. 1.11, fourth figure on the top left). Rotate the grille once more and write the next 4 letters in the empty cells that appear through the grille's windows (Fig. 1.11, first figure on the bottom left). Finally, the last 90° rotation exposes the last 4 empty cells in the table into which the last 4 letters of the message are written. When the grille is lifted, the table is now filled with all the letters of the message, jumbled up. Only those with the rotating grille can decipher the message (by placing it on the table and rotating it to expose all the letters).

The history of codes and ciphers has been discussed in many books and publications. While substitution and transposition ciphers are historically significant, modern cryptographic methods have been developed and used daily anywhere encryption is needed.

Generalisations and Further Explorations

The puzzle at the beginning of this chapter was a deciphering puzzle. Given ciphertext and a key, you were asked to decipher the code. The next step up is to try and break a code, given the ciphertext, but without the key. Breaking codes makes for great puzzles. Of course, knowing what kind of cipher encoded the message is necessary so that we can apply an appropriate cryptanalytic method. But there is something we can do even if we don't know anything about the cipher. Take a look at the following two ciphertexts.

1. QZIOGMEBTXVJBIEBUQZIWGGEIUKZITQXVJBIEBUQZITFUEQZJI IZVUEJIEPIQIJTWITQGNQZIRFMPQJII

2. ERWSNHEEADIERSHTFRTBHHDDDSEOEOBHEMTMHUEEITRW PLDDUEEEOTGIOTTRDSLINNIAUETESTCENNRHE

One of them uses a substitution cipher; the other is a transposition. You can tell which is which quite easily. The frequency of letters in the transposition is similar to that of the English language.

Since the message in line 2 has many E's and T's — two of the most frequent letters in the English alphabet, we deduce that it is the transposition. Moreover, not only can we infer that ciphertext number 1 is the substitution, but it is reasonable enough that Q and I substitute T and E, respectively, since they are very frequent in the ciphertext. We also might safely suggest that the triplet QZI, which appears four times in the sentence, including as the first word in the text, encodes the word "The". Once we know the three letters T, H, and E and plug them in, we can continue to "break" the substitution cipher by guessing words. Guessing, trial and error methods, and even "gut feelings" are often employed by cryptanalysts, as testified by many of the codebreakers in Britain's code-breaking headquarters at Bletchley Park during the Second World War. Admittedly, when there are no spaces between the words, this makes things slightly more difficult, but not too hard for a good codebreaker. You can try and finish off the code breaking on ciphertext 1. Don't worry. I'll give you the answer on the following page.

If you know the size of the table used to encode a regular columnar transposition cipher, there's an easy way to break the code, which we'll get to shortly. Sometimes, however, the size of the table is unknown, so you have to try and deduce it. This means finding all the different tables that will hold the given ciphertext. How difficult is this? This depends on how many ways there are to *factor* the number of letters in the ciphertext into two *factors*. A *factor* of a given number is a (positive) whole number that divides the given number without a remainder; for example, 1, 2, 3, 4, 6, and 12 are all factors of 12. For all practical reasons, we can find the answer for small numbers using one of the many algorithms that have been developed to solve this problem. However, for very large numbers, this problem is an open problem in mathematics. Mathematicians have not yet found a formula that can tell us the answer or even an algorithm that can find the factors in a reasonable amount of (computer) time.

Let's take a look at ciphertext number 2. There are 81 letters in the text. We can deduce from this that the size of the table can either be 9×9 or 3×27. 9×9 seems more likely, so we'll go with that and divide the ciphertext into chunks of nine letters like this:

ERWSNHEEA   DIERSHTFR   TBHHDDDSE   OEOBHEMTM
HUEEITRWP   LDDUEEEOT   GIOTTRDSL   INNIAUETE
STCENNRHE

We then write each of the nine 9-letter groups in columns on strips of paper or a spreadsheet (Fig. 1.12).

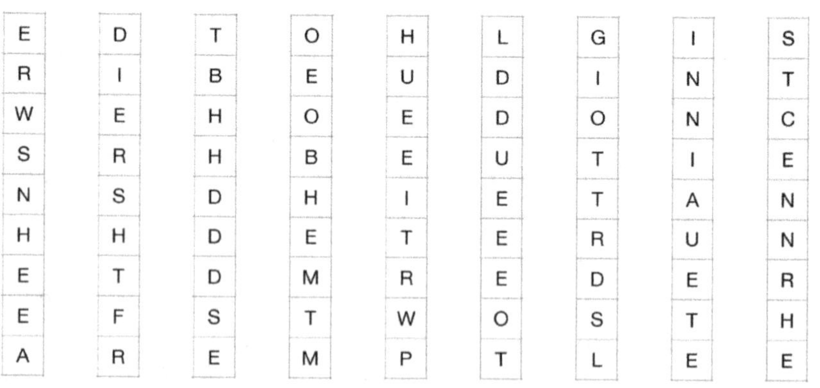

Fig. 1.12   Columns of ciphertext number 2

Now, try and rearrange the columns in the correct order by placing two columns next to each other if that yields sensible pairs of letters. For example, it would be reasonable to argue that the fifth column, beginning with an H, comes after the column beginning with a T and before the column beginning with an E to form the word THE at the beginning of the plaintext. It would also seem less plausible that the last column, which has an S in its first row, would come before the seventh column, which has a G in its first row. SG just doesn't make a good pair. Neither do NR or HS, the pairs in the sixth and eighth rows, respectively. Of course, you have to be careful since this *can* happen if the first letter in the suggested pair is the end of one word and the second is the beginning of another. So it might just be, for example, that the three aforementioned pairs are ends and beginnings of phrases like: "i**S G**ood", "i**N R**easonable", or "wit**H S**afe", but this is less likely to happen in a third of the rows in the column. Once again, statistics play an important part in cryptography. In fact, computer programs that crack columnar transposition ciphers use statistics to throw together columns whose pair frequencies, down all the rows, are favourable. If the message is short, they can use "brute force". This means that all the possible combinations of column positions are tried until a plaintext message appears. The computer doesn't have to show us all the possible combinations. It uses statistics once more to test if the resulting text is plaintext or just a garbled message.

The number of possible combinations for a given number of columns can be calculated using *combinatorics*, an area in math that deals with, well, counting things! Let's try and heuristically find the number of possible combinations for a nine-column transposition cipher, like ciphertext number 2. One way to do this is to work out the math for a simpler problem and then extrapolate until we get a general mathematical formula. Suppose there are just two columns in a columnar transposition cipher. Obviously, there are two ways to arrange them, one before the other. If we label the columns *a* and *b*, we can write down the two different arrangements as *ab* and *ba*. If there are three, let's label them *a*, *b*, and *c*. The six possible arrangements are: *abc, acb, bac, bca, cab, cba*. It's easy to see how the two arrangements for the two-column cipher become six for the three-column cipher. The additional column, labelled *c* can be put in three different places in each of the two arrangements: before the first column, between the first and last column, or after the last column. Since $3 \times 2 = 6$, there are six different arrangements for the three-column case. Using the same logic, we can deduce that with four columns, there are $4 \times 3 \times 2 = 24$ arrangements. If you are not convinced, label them and write them out for yourselves. Can you see the emerging pattern? For five columns, there are $5 \times 4 \times 3 \times 2 = 120$ possible arrangements; for six, $6 \times 5 \times 4 \times 3 \times 2 = 720$; for seven, $7 \times 6 \times 5 \times 4 \times 3 \times 2 = 5040$; for eight, $8 \times 7 \times 6 \times 5 \times 4 \times 3 \times 2 = 40320$; and for nine, $9 \times 8 \times 7 \times 6 \times 5 \times 4 \times 3 \times 2 = 362880$ different arrangements. By the way, calculating the product of all the whole numbers up to a given number is called taking the *factorial* of the (given) number. The *factorial* is an important mathematical operator and is represented by placing an exclamation mark after the number. So, I could have shortened the previous sentences by writing: 3! = 6, 4! = 24, 5! = 120, and so on.

Trying all the 9! arrangements by hand would take a *long* time. That's why using statistics and heuristics is so important. True, a computer *can* use brute force, but the number of different arrangements for even a message of moderate size, say, a message that fits a 32×32 grid (1,024 letters, approximately 218 English words), becomes a humongous 32!, which is a number that is 36 digits long (2.63×10^{35}), way beyond the capabilities of computers today.

Back to ciphertext number 2. I urge you to try and break the code yourself before reading on. The solution is given in Fig. 1.13. In fact, both ciphertexts 1 and 2 encode the following message: "The gold is buried in the wooden chest buried in the sand three hundred meters west of the palm tree."

T	H	E	G	O	L	D	I	S
B	U	R	I	E	D	I	N	T
H	E	W	O	O	D	E	N	C
H	E	S	T	B	U	R	I	E
D	I	N	T	H	E	S	A	N
D	T	H	R	E	E	H	U	N
D	R	E	D	M	E	T	E	R
S	W	E	S	T	O	F	T	H
E	P	A	L	M	T	R	E	E

Fig. 1.13   Solution to ciphertext number 2

Cryptographers, intelligence agencies, and puzzlers have already generalised columnar transposition ciphers in many ways. We've mentioned some of them: adding "windows" in the table is one example. The rail fence and route ciphers are others. Generalising the idea of these, we can suggest either changing the shape of the table (Fig. 1.14) or changing the shape of the route on the table (Fig. 1.15) to get similar ciphers of this kind. Once the shape of the table is changed, like the one in Fig. 1.14, the ciphertext can be

Fig. 1.14   A hexagonal transposition cipher

T	I	P	T	I	E	K	E
P	H	O	E	C	E	A	S
N	H	N	L	T	E	N	S
E	U	R	I	D	A	O	R
E	T	O	R	E	M	F	R
S	R	D	U	M	B	E	F
I	I	A	R	M	S	R	O
E	E	N	S	A	R	A	P

Fig. 1.15   Table with plaintext deciphered by the route specified in Fig. 1.16

extracted in any way you like. If it is taken in a vertical route from top to bottom, the ciphertext is: NHTIWTINTHMILARTOUMTLE?CE HRAUEEF.

The ciphertext in the route cipher in Fig. 1.15 can be extracted either down columns or along rows. Since the plaintext is recovered along a route, there is no benefit in using a key or extracting ciphertext in any other way.

Figure 1.16 shows the route the text is extracted, which is depicted using both numbers and coloured lines. Starting from the cell labelled 1 in the top left corner and then going to the cells labelled 2, 3, and so on, the extracted plaintext reads: "There is desperate need for more supplies of ammunition in the rear barracks."

The route in Fig. 1.16 is especially interesting. The 8×8 table can be viewed as a chessboard. Note that the route consists entirely of consecutive knight moves in the game of chess. For those of you who are not familiar with chess, the knight, one of the pieces in the game of chess, moves by going two squares in one direction (up, down, right, or left) and then one square in one of the perpendicular

Fig. 1.16   Route with which to extract the plaintext letters from the table in Fig. 1.15

directions, or vice versa (one square in one direction and then two squares in the perpendicular direction). Follow the route on the chessboard in Fig. 1.16 and convince yourself that it is indeed a so-called *knight's tour* of the chessboard, visiting each square exactly once. Finding knight tours on different kinds of tables is an interesting, elementary problem in recreational maths and computer science. Indeed, the number of possible tours on many boards, including chessboards larger than 8×8, is still unknown and a regular challenge for mathematicians and computer scientists. The knight's tour is said to have originated in the Islamic Empire during the 9th century A.D., possibly even earlier, and has been utilised as a puzzle ever since. It is interesting to note that in the 19th century, people used to create puzzles by chopping up well-known poems into words and then placing the words one at a time into a table according to a knight's tour. These puzzles, also called cryptotours, resemble the concept of route ciphers, using words instead of numbers. Another knight tour pastime is called knight tour art, and it's all about creating cool patterns and images using knight tour algorithms. Recreational mathematician George Jelliss has a whole section on knight's tours on his excellent website, mayhematics, including examples of tours on different kinds of boards, tours with other chess pieces, and lots of history and theory.

Grille ciphers can also be generalised to shapes other than squares, but to do this with *rotating* grilles, we need to understand how to place the windows on the cardboard or paper key that we put over the table with plaintext. This has to be done carefully so that the windows cover all the possible places on the board without any two windows overlapping any cell. Here's how it can be done with a square, cardboard, or paper grid. Divide the grid into a series of nested, square rings, and number them as shown in the 6×6 example on the left-hand side of Fig. 1.17. You create the windows by cutting out one of each number in each ring. In Fig. 1.17, for example, you cut out one 1, one 2, one 3, one 4, and one 5 in the outer ring; one 1, one 2, and one 3 in the middle ring; and one 1 in the inner ring as shown in the grid on the right-hand side. You now have the key to place over the table with the ciphertext. Note that by employing this method, you cut out exactly a quarter of the number of cells on the grid and that each time you rotate the grid by 90°, the windows uncover cells that have not previously been exposed. If you have a square, odd number of cells, the method is exactly the same, except that the centre cell cannot be used. This method can be used for other shapes as well, provided they are regular; that is, their sides are all equal, and so are their inner angles. Figure 1.18 is an example of a hexagonal grille, just before marking the windows.

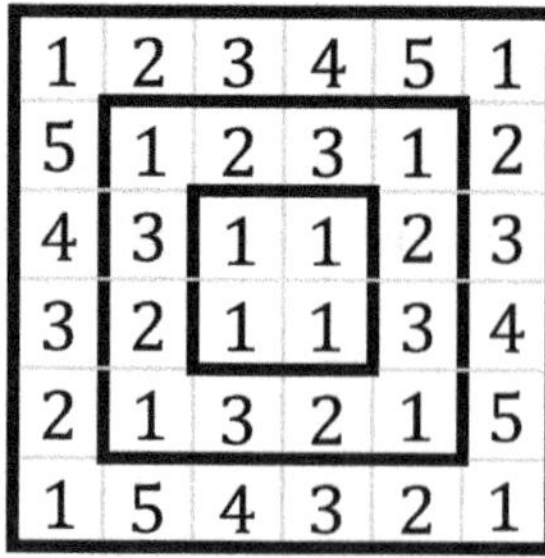

Fig. 1.17   How to create a rotating grille

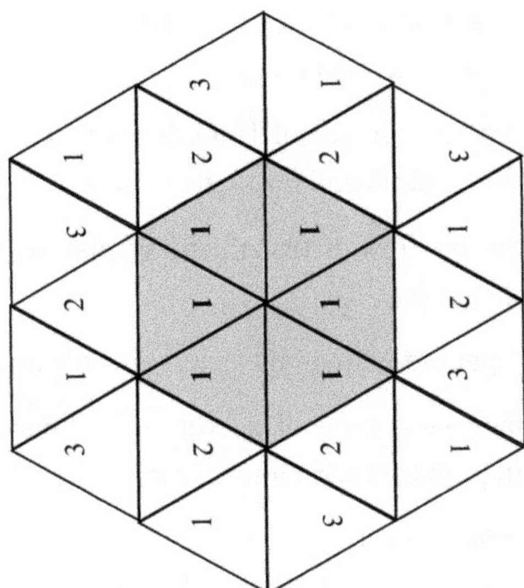

Fig. 1.18   A rotating hexagonal grille

Recap

Codes and ciphers have been around for a long time. At the beginning of the chapter, we took a short look at some substitution ciphers and how to break them, but we focused mainly on transposition ciphers. We learned about the connection between cryptography and the mathematical areas of combinatorics and statistics and discussed some of the different types of transposition ciphers and some possible generalisations. It's important to understand that substitution and transposition ciphers are only two of a whole genre of cryptographic methods, and I urge those of you who are interested to read some of the stuff in the reference section to learn more about this fascinating subject.

Here are a few short, informal definitions of some of the concepts we encountered in this chapter:

- *Plaintext* — plaintext. A word, sentence, or entire message in legible language.

- *Ciphertext* — ciphertext. A word, sentence, or entire message that has been encoded so that it is illegible.

- *Cipher key* — the rule or set of rules by which the plaintext is transformed into the ciphertext and vice versa.

- *Encryption* — the process of transforming plaintext into ciphertext, using a cipher key.

- *Cryptanalysts* — people who analyse codes to try and break them.

- *Substitution cipher* — a type of secret code where you replace each character in plaintext with another letter, number, or symbol.

- *Transposition cipher* — a type of secret code where you convert plaintext into ciphertext by jumbling the characters according to some function (the key).

- *At-bash* — a substitution encryption method where you replace the first letter of the alphabet with the last letter of the alphabet, the second letter with the second to last, the third with the third to last, and so on.

- *Caesar cipher* — a substitution encryption method where you replace each letter with a letter a constant distance away from it in the alphabet.

- *Columnar transposition cipher* — a transposition cipher where you write the plaintext letters in the cells of a rectangular grid, in order, along the rows, and then create the ciphertext by extracting the letters down the columns in a pre-determined order (the key).

- *Double columnar transposition cipher* — a columnar transposition cipher where, after extracting the ciphertext, it is written in the table along the rows and extracted once more according to the given key.

- *Scytale* — a transposition cipher, where to encrypt a secret message, you wrap a strip of leather in a spiral around a rod, write the plaintext letters horizontally on the wrapped leather, and extract the ciphertext by unwrapping the leather strip and reading the letters down the unwrapped strip.

- *Rail fence cipher* — a transposition cipher where plaintext is written in zigzag down and up the columns of a rectangular table, and the ciphertext is extracted horizontally.

- *Route cipher* — a transposition cipher where the plaintext is written according to a pre-determined route through the cells in a grid, and the ciphertext is extracted in another way, usually along the rows or down the columns.

- *The Myszkowski Transposition* — a columnar transposition cipher where at least two columns are numbered with the same number, and the letters from these columns are extracted as adjacent pairs.

- *Grille cipher* — a transposition cipher where, given a rectangular grid covered by a piece of paper or cardboard the same size with holes or windows in it (the grille), letters of a message are written through the windows onto cells on the grid underneath whereupon the grille is removed, fake letters are added to the grid, and the ciphertext is extracted in a pre-determined way, usually along the rows or down the columns.

- *Rotating grille cipher* — a grille cipher with a specially designed grille so that after it is placed initially over the grid and the plaintext letters are written through the windows, it can be rotated a few times (according to the symmetry of the grid), where each time you can write some more plaintext letters through the windows.

- *Hieroglyphs* — a form of symbolic writing used in Ancient Egypt that used symbols or pictures to represent words or sounds.

- *Factors* — the (positive) whole numbers that divide a given number without a remainder.

- *Factorial* — a mathematical operator, denoted by an exclamation mark that multiplies all the whole numbers up to the given number.

- *Combinatorics* — an area in math that studies ways of counting things.

- *Knight tour* — a series of knight moves on a chessboard where the knight visits all the squares on the board exactly once.

- *Cryptotour* — a route cipher on a chessboard whose key is a knight's tour.

Challenge Yourself!

(1) The following ciphertext has been encoded using a Caesar shift substitution cipher. Break the code and decipher the message to reveal a paragraph from the Wizard of Oz!

> "Itwtymd ymtzlmy xmj btzqi lt sjcy; xt xmj yttp Ytyt ns mjw fwrx fsi hqnrgji ts ymj Qnts'x gfhp, mtqinsl ynlmyqd yt mnx rfsj bnym tsj mfsi. Ymj sjcy rtrjsy ny xjjrji fx nk xmj bjwj kqdnsl ymwtzlm ymj fnw; fsi ymjs, gjktwj xmj mfi ynrj yt ymnsp fgtzy ny, xmj bfx xfkj ts ymj tymjw xnij. Ymj Qnts bjsy gfhp f ymnwi ynrj fsi lty ymj Yns Bttirfs, fsi ymjs ymjd fqq xfy itbs ktw f kjb rtrjsyx yt lnaj ymj gjfxy f hmfshj yt wjxy, ktw mnx lwjfy qjfux mfi rfij mnx gwjfym xmtwy, fsi mj ufsyji qnpj f gnl itl ymfy mfx gjjs wzssnsl ytt qtsl."

(2) The following ciphertext encodes a poem taken from *Alice in Wonderland*. This cipher is a simple 10-column transposition code. Break the code to reveal the poem!

hlovnoreloloasnlengoidegusodwygthdistjwtihtronechrliwt
iladtliatfenheiysetnywolesihtvsesnscllwssteisleheceehplcei
mxhcmhawerareoraoftixtrpinanylfmwewmihlxhorndtigeu
snasesgixecoipelohltedashenx

(3) Break the following 3-line rail fence cipher! Hint: Write the first line with 3 spaces between every two letters, the second line with 1 space between every two letters, and finish the bottom rail of the zigzag in the third line. It might help to write the ciphertext in a 3-line, 67-row table.

AEBNEEEAEEIRETTELSHSNEATSTVRHOENHWVSHSEESCE
SOHSAULTUGOOTCTAWPDRTEGS

(4) Break the following hexagonal transposition cipher using the same hexagonal grid as in Fig. 1.14; the grid has seven lines, the first line 2 letters long, the second, 4, the third, 6, the fourth, 8, the fifth, 6, the sixth, 4, and the last line, 2 letters long. The plaintext is read along the rows; however, the method of extraction of the cipher from the grid is different. Hint: The cipher is extracted in a zigzag fashion up and down the columns.

INASWINOGUERSNARSTTHWNRTEIHEEPGI

(5) Decode the following route transposition cipher using the same table as Fig. 1.15 and the same route as Fig. 1.16. This time, the ciphertext is written in the table in order along the rows instead of down the columns.

TEHTTRGP   TITEDOCH   RHAAETEE   NTIERIEN
NRGTLMTR   SGLYUHEE   STOGEIBS   TEIEESNG

(6) Apply the rotating grille on the right-hand side of Fig. 1.19 to the cipher table on the left-hand side to recover the plaintext. You can do this by printing out the grille and the table, cutting out the black cells in the grid and then placing it on the table,

reading out the letters that appear in the grille's windows, and then rotating clockwise and repeating until you've uncovered all the plaintext.

Fig. 1.19 Rotating grille problem

(7) Apply the rotating hexagonal grille on the right-hand side of Fig. 1.20 to the cipher table on the left-hand side to recover the plaintext.

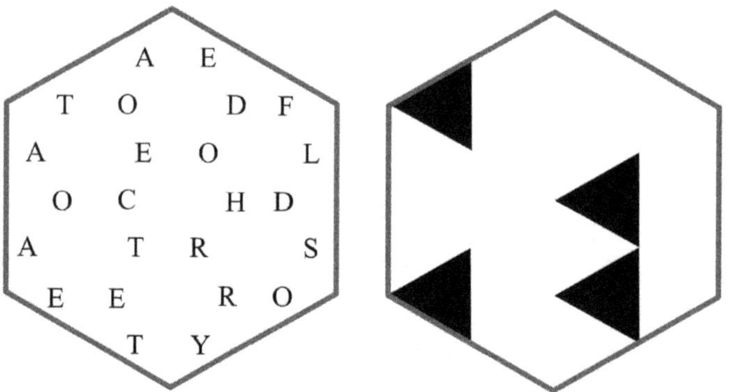

Fig. 1.20 Rotating hexagonal grille problem

Solutions

(1) The cipher is a shift 5 Caeser substitution (a in ciphertext = f in plaintext, b in ciphertext = g in plaintext, c in ciphertext = h in plaintext, and so on).

The paragraph reads:

"Dorothy thought she would go next; so she took Toto in her arms and climbed on the Lion's back, holding tightly to his mane with one hand. The next moment it seemed as if she were flying through the air; and then, before she had time to think about it, she was safe on the other side. The Lion went back a third time and got the Tin Woodman, and then they all sat down for a few moments to give the beast a chance to rest, for his great leaps had made his breath short, and he panted like a big dog that has been running too long."

(2) The poem is:

"How doth the little crocodile
Improve his shining tail,
And pour the waters of the Nile
On every golden scale!

How cheerfully he seems to grin,
How neatly spreads his claws,
And welcomes little fishes in,
With gently smiling jaws!"

Fig. 1.21   The head of the little crocodile

And the columnar transposition table is:

1	2	3	4	5	6	7	8	9	10
h	o	w	d	o	t	h	t	h	e
l	i	t	t	l	e	c	r	o	c
o	d	i	l	e	i	m	p	r	o
v	e	h	i	s	s	h	i	n	i
n	g	t	a	i	l	a	n	d	p
o	u	r	t	h	e	w	a	t	e
r	s	o	f	t	h	e	n	i	l
e	o	n	e	v	e	r	y	g	o
l	d	e	n	s	c	a	l	e	h
o	w	c	h	e	e	r	f	u	l
l	y	h	e	s	e	e	m	s	t
o	g	r	i	n	h	o	w	n	e
a	t	l	y	s	p	r	e	a	d
s	h	i	s	c	l	a	w	s	a
n	d	w	e	l	c	o	m	e	s
l	i	t	t	l	e	f	i	s	h
e	s	i	n	w	i	t	h	g	e
n	t	l	y	s	m	i	l	i	n
g	j	a	w	s	x	x	x	x	x

Fig. 1.22   Solution to the columnar transposition code problem

(3)

```
A   E   B   N   E   E   E   A   E   E   I   R   E   T   T   E   L
 S H S N E A T S T V R H O E N H W V S H S E E S C E S O H S A U L 
  T   U   G   O   O   T   C   T   A   W   P   D   R   T   E   G   S
```

Fig. 1.23   Solution to the rail-fence problem

(4)

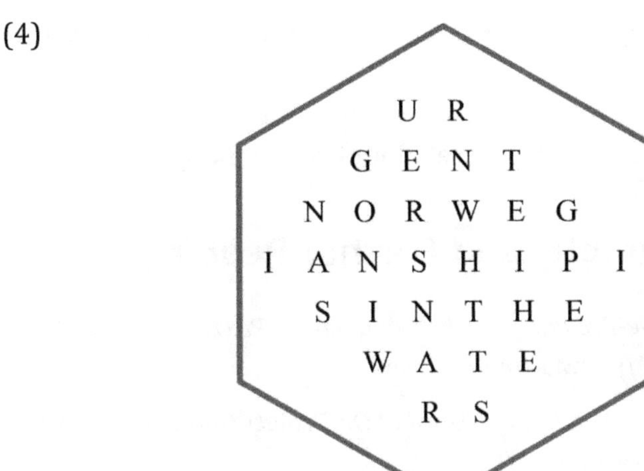

Fig. 1.24   Solution to the hexagonal cipher problem

(5) The cipher table with the ciphertext is shown in Fig. 1.25. The
plaintext reads: "The intelligence reports suggest that the enemy
is retreating to the bridge."

T	E	H	T	T	R	G	P
T	I	T	E	D	O	C	H
R	H	A	A	E	T	E	E
N	T	I	E	R	I	E	N
N	R	G	T	L	M	T	R
S	G	L	Y	U	H	E	E
S	T	O	G	E	I	B	S
T	E	I	E	E	S	N	G

Fig. 1.25   Correct cipher table for the route cipher problem

(6) The plaintext reads: "There are many ways to encode a secret email."

(7) The plaintext reads: "There are a lot of codes today."

Bibliography and Further Reading

Åhlén, J. (2023). *Code-Breaking, Cipher and Logic Puzzles Solving Tools.* Boxentriq. https://www.boxentriq.com/

Baum, L. F. (1993). *The Wonderful Wizard of Oz.* Project Gutenberg. https://www.gutenberg.org/ebooks/55

Carroll, L. (2009). *Alice's Adventures in Wonderland.* Project Gutenberg. https://www.gutenberg.org/ebooks/28885

Elran, Y. (2016). *Maths Puzzles: Cryptarithms, Symbologies and Secret Codes [MOOC].* Future Learn. https://www.futurelearn.com/courses/maths-puzzles

Elran, Y. (2020). *Lewis Carroll's Cats and Rats ... and Other Puzzles with Interesting Tails.* Singapore: World Scientific.

Gardner, M. (2013). *Codes, Ciphers and Secret Writing.* United States: Dover Publications.

Jelliss, J. (2023). *Knight's Tour Notes.* Mayhematics. http://www.mayhematics.com/t/t.htm

Sigmon, N. P. and Klima, R. E. (2012). *Cryptology: Classical and Modern with Maplets.* United Kingdom: Taylor & Francis.

Singh, S. (2000). *The Code Book: The Secret History of Codes and Code-breaking.* Fourth Estate.

Chapter 2

Lewis Carroll's Doublet Puzzles and Mathematical Poetry

Introduction

Many mathematicians dabble in word games. In fact, there's a lot of math lurking amongst some of these puzzles. We begin with doublets, one of my favourite word puzzles, invented by Charles Dodgson, aka Lewis Carroll, mathematician, logician, and the author of the "Alice in Wonderland" books.

The Puzzle

Doublets is a puzzle where you turn one word into another in a number of steps, using the following rules:

- *In each step, replace one of the letters with any other as long as you get a genuine English word.*

- *You are allowed to use different tenses, prefixes, suffixes, etc.*

- *You are **not** allowed to rearrange the order of the letters in the word at any stage.*

- *Use the minimal number of steps possible.*

Here is an example. Turn CAT into DOG in the least number of steps.

- **CAT**

- *COT*

- *DOT*

- **DOG**

The number of steps (3) used in this example is the smallest possible. Here are a few doublets for you to solve!

1. *Turn RUST into GOLD in less than six steps.*

2. *Turn FAST into SLOW in less than seven steps.*

3. *Turn PINK into BLUE in less than eleven steps.*

In some cases, there is no way to verify that the number of steps is minimal — so if you find a solution with fewer steps than I did, please let me know!

Where to Start?

Trial and error is the standard way to solve doublets; however, some tips might be useful. The first thing to do is systematically replace one of the letters with the corresponding letter in the target word and then check if the outcome is valid. Sometimes, none of the letters from the target word can be used, and then an interim word has to be chosen. In this case, try and choose a word that, in the *following* step, can be transformed into a new word using a letter from the target word. Sometimes, it's easier to work backwards — from the target word to the beginning.

Solving the Puzzle

Here are possible solutions to the puzzle:

- ***RUST***
- *MUST*
- *MOST*
- *MOLT*
- *MOLD*
- ***GOLD***

- ***FAST***
- *FART (Ooops…)*
- *FORT*
- *FOOT*
- *SOOT*
- *SLOT*
- ***SLOW***

- ***PINK***
- *PICK*
- *POCK*
- *PORK*
- *PORT*
- *POUT*
- *POUR*
- *SOUR*
- *SLUR*
- *BLUR*
- ***BLUE***

The History of the Puzzle and its Inventor

Lewis Carroll (Charles Lutwidge Dodgson) (Fig. 2.1) was born on January 27, 1832, in Daresbury, England. His father, also a Charles Dodgson, was gifted in math but worked as a country parson, in line with many of his ancestors. Carroll studied math at Oxford. He became a lecturer at the university and published several works on mathematics and logic; however, it was his literary works, especially "Alice in Wonderland" and "Through the Looking Glass", which earned him lasting fame. These began as stories that Carroll made up for the three daughters of Henry Liddell, the dean of Christ Church, during rowing trips on the Thames that he used to take them on. Carroll's "Alice" books are full of word games, but so are Carroll's many other writings including 15 literary works, 13 mathematical works, and many papers, letters, and articles in magazines of the time.

"Doublets" gained fame as a puzzle when Carroll published it in the March 1879 edition of *Vanity Fair*, a popular British weekly

Fig. 2.1   Photograph of Lewis Carroll from 1863

magazine published from 1868 to 1914 (not to be confused with the American publication with the same name). Doublets is just one example of Lewis Carroll's obsession with wordplay. Another is *anagrams* — jumbling up the letters in a word (or sentence, paragraph, etc.) to get a new one. If they're entertaining, like "debit card" and "bad credit", or synonymous, like "astronomer" and "moon starer", all the better!

Carroll's literary genius can be seen in almost every piece of writing he wrote. Take a look at his poem "How doth the little crocodile":

> *"How doth the little crocodile*
> *   Improve his shining tail*
> *And pour the waters of the Nile*
> *   On every golden scale!*
>
> *How cheerfully he seems to grin,*
> *   How neatly spreads his claws,*
> *And welcomes little fishes in*
> *   With gently smiling jaws"*

Compare this with Isaac Watt's "Against idleness and mischief":

> *"How doth the little busy bee*
> *   Improve each shining hour,*
> *And gather honey all the day*
> *   From every opening flower!*
>
> *How skilfully she builds her cell!*
> *   How neat she spreads the wax!*
> *And labours hard to store it well*
> *   With the sweet food she makes.*
>
> *In works of labour or of skill*
> *   I would be busy too:*
> *For Satan finds some mischief still*
> *   For idle hands to do.*

> *In books, or work, or healthful play*
> *Let my first years be past,*
> *That I may give for every day*
> *Some good account at last."*

Carroll's poem is obviously a spoof on Watts's. It captures Carroll's rebellious stand against the haughty Victorian self-righteousness. The poem appears in chapter 2 of "Alice in Wonderland". Alice is at the beginning of her adventures, worried that she can't properly recite "Against idleness and mischief", a poem she was obviously supposed to know by heart. So, "How doth the little crocodile" is her failed attempt, and after the first two stanzas, she gives up. Carroll's nonsensical poem, in my mind, is a hallmark of intelligence. His mimicry of Watts is faultless — the same stance structure, the same iambic rhythm, and the same words at the beginning of each line. Many of the words even sound the same — "every" and "gently", or "neatly" and "neat she". The juxtapositions were carefully chosen. The bee is as diligent as the crocodile is lazy. Honey is gathered while water is poured. What a great poem!

Lewis Carroll's wordplay sparked so much interest that it eventually evolved into a genre in its own right, known today as "recreational linguistics". In particular, doublets has been combined with the popular pattern-matching puzzle "Mastermind" appearing very recently as Wordle™. We'll learn about this and other "bits and bobs" in the next section. Lewis Carroll died from influenza on January 14, 1898, but his legacy today remains as relevant as it was a hundred years ago.

Generalisations and Further Explorations

When I challenge kids in my math classes with some of Lewis Carroll's doublets or discuss with them some of Alice in Wonderland's capers, I often get asked: "What has this to do with math?" I believe that math is language, and language is math; there is math about literature and literature about math. I'll try and explain what I mean in the next few paragraphs.

What is a language? Language expresses concepts by turning them into discrete symbols, words, or phrases that can be communicated to others. These concepts are complex mental constructions we use to organise and understand our experiences. Language reduces them quite a lot from their original meaning because it necessarily involves a process of simplification and abstraction according to rules and conventions, prioritising clarity, concision, and simplicity. In other words, we can't really grasp the "full meaning" of anything; there will always be something "lost in translation". Take the concept of "dog", for example. The minute you read the word "dog", each and every one of you reduced it to your own individual representation of a "dog". The dog I am thinking of right now is the small, black cocker spaniel called "Poppy", with whom I grew up as a kid. Did any of you think of "Poppy"? I doubt it. The concept "dog" constitutes the collection of all the individual interpretations from all times, all over the world, including many interpretations that are not even man's four-legged best friend, for example, the meaning of the word "dog" in the sentence "You can't teach an old dog new tricks". Other concepts are even more difficult to capture. If we switch the letters G and D in the word dog, we get god. Try and explain that concept!

What else can we say about language? Language is inherently subjective and culturally constructed. Different languages and cultures have different ways of conceptualising and expressing ideas, which can lead to further reduction and simplification of the original concept. Certain concepts in one language may not have direct translations in another. Some words are, therefore, just incorporated as is into another language, like the Yiddish schmooze or the Italian pizza. I recently heard a game show on the radio where the host took an English word, translated it into Japanese using an online translator, translated the answer into Turkish, and then translated it back to English. The guests on the show were given the triple-translated word and had to guess the original. Boy, was that difficult! If you do this with the word "ice" using Google Translate, you get "curry". Go figure!

The way concepts are rendered as language can also be different. Far Eastern languages are often pictorial, where one symbol represents the concept. Others are based on a small set of letters that can be fused into words, sentences, paragraphs, etc.

What about math? Is mathematical notation a language, so to speak? Is notation reductive (like language), or does it capture the concept entirely? Is math also "inherently subjective and culturally constructed"?

Perhaps mathematical notation is the "ultimate language", one that is universal and objective and comes closest to the original meaning of the concepts it tries to convey. Linguists, philosophers and mathematicians all argue about the answers to these questions. I, for one, certainly advocate regarding math as a language, and this has important repercussions on how I think it should be taught at school. In most schools, math is not taught or learned as if it were a language, which I think is the primary reason many people fear math. This is totally understandable. I still get intimidated when I look at a math paper full of symbols I've never seen before, e.g., words in Latin like *Lemma* and *Theorem*, and then read sentences beginning with "It's trivial to see that..." In order to learn math as a language, the concepts must be learned, practised, and experienced in real-world contexts. The student has to be "immersed" in math, and this should start at a very early age.

We now turn to the interconnectivity between math and literature. Mathematics and literature may seem very different, but they share surprising similarities. The obvious connection is when someone writes about math, like the book you are reading now. Another is when mathematical ideas are explicitly or implicitly included within a text. For example, in the classic novel *Flatland* by Edwin Abbott, the characters live in a two-dimensional world explored using mathematical concepts in geometry and topology. Similarly, in the popular book *The Da Vinci Code* by Dan Brown, the plot is heavily influenced by mathematical ideas from cryptography and code-breaking.

Another important junction where math and literature meet is math poetry. Romanian-born mathematician and math poet Sarah Glaz wrote an excellent historical review of math poetry in the *Journal of Mathematics and the Arts*. She points to "The Herds of Nanna", an ancient Sumerian temple hymn inscribed on a clay tablet around 1800 B.C. to help count the ever-increasing amounts of sacrifices, as one of the first math poems. A thousand and six hundred years later, Archimedes, the famed Greek mathematician and physicist, wrote a "sacrifice counting" problem in poetic form. His poem, however, is far more sophisticated. Known as "The cattle problem", it asks for the number of cows and bulls of four different colours, given certain restrictions. Beginning with the words: "If thou art diligent and wise, O stranger, compute the number of cattle of the Sun, who once upon a time grazed on the fields of the Thrinacian isle of Sicily, divided into four herds of different colours...", the poem invites the reader to solve two problems — which you'll find in Chapter 6.

In Archimedes' cattle problem, the content of the poem is mathematical. Sometimes, however, there is also math in the *structure* of literary works. Poetry, for example, often follows a prescribed format in the same way that mathematical equations follow strict rules and patterns. In Haiku, a popular 3-line form of poetry, the structure is: the first and last lines have five syllables, and the middle line has seven. The Malayan Pantoum is a poem comprising four-line stanzas, where the second and fourth lines in each stanza are repeated as the first and third lines of the next stanza, respectively. *The Teachers and Writers Handbook of Poetic Forms*, edited by Ron Padgett, is an excellent source for people who want to understand the underlying structure of poetry.

I have found that writing Haikus is a great way to engage kids in math lessons! Here is a Haiku I came up with:

"One plus one is two,
Two plus two is twenty-two,
No, that's wrong! It's four."

... and here is one that the generative artificial intelligence bot ChatGPT suggested when I prompted it with, "Write a Haiku on a math topic!":

"Infinity realm,
Mysteries beyond our grasp,
Mathematics soars."

Which do you think is better?

I was lucky enough to learn a lot about math poetry at a "creative writing in maths" workshop that I attended in Banff, Canada. The workshop was arranged by Marjorie Senechal, a mathematician, historian of science, playwright, and author, who was for many years the editor-in-chief of *The Mathematical Intelligencer*, a math journal whose mission is "to inform, to entertain, and to provoke". One of the participants I met at the workshop was developmental psycholinguist and poet Robin Chapman, who has written ten full-length poetry collections and won many awards and prizes. Here is her poem, "Distance, Rate, Time":

Distance, Rate, Time

"Going fast
Down the path,
Fighting for breath,
For footing,
The run becomes

One long pouring
Of legs, lung, salt

Into green, gravel, wind,
Green legs, gravel lungs,
Salt wind.

While time,
Hawk on the updraft,
Climbs overhead,
Turning and turning."

Robin told me this poem was conceived while thinking about the equation for distance, *distance = rate × time*, as she went on her daily runs. She found herself thinking of life history as well as running time — and then was surprised by the coincidence of the hawk and metaphor at the end. In her own words: "Of course, that's what equations permit as well as metaphors — a vehicle for the meanings — mortality, life span — that can be assigned!"

Another scientist-come-poet at the workshop was Madhur Anand, an ecology and environmental science professor at Guelph University. Madhur has also won awards for her poems and is considered one of the founders of the sub-genre of eco-poetry, poetry that puts a strong emphasis on ecological concerns. One of her most creative ideas is to write poems composed solely from words and phrases from her scientific articles. Such is her poem, "Especially in Time":

Especially in Time

"Wild populations recognize that the linearity,
the relative rareness, the major museums, or any area
which is known, is a surrogate
for proximity

Stream beetles, Galapagos finches, and Israeli
passerine birds are transformed
into an index of limited
available information

Elytral lengths, slope of the regression,
and mid-latitude precipitation
unravel the anomalies

A prolonged change is also under scrutiny."

If you are interested in the scientific article from which all the words in the poem are taken, you'll find it in the bibliography section.

Another favourite of mine is Madhur's poem, "Moving On", particularly because in her book, the page begins with two quotes, one from Lewis Carroll's *Through the Looking Glass* and the other from a meeting regarding a Ph.D. thesis.

Moving On

It takes all the running you can do, to keep in the same place.

    – Lewis Carroll, *Through the Looking Glass*

Temporal stability of sites estimated by a modified, reciprocal coefficient of variation.

    – From a Ph.D.-in-progress supervisory
      committee meeting (2014)

"I'm on a stationary bike looking at numbers.
  Heart rate, calories consumed, distance travelled,
   time spent,
  instantaneous speed. Each motivates, differently.
  Health. Weight. Destination. When will I be done
   with this?
  How am I doing? I've been obsessed with
   averages.

But now small fluctuations of speed don't
  correlate

> *exactly with my heart. Variation! How neurons*
> *fire with this song. How the mind goes places.*
>   *How one thigh*
> *takes the lead, so cycles are never perfect. How*
>   *with time*
> *small deviations accumulate from sensitive*
> *dependence on initial conditions. Then chaos.*
> *By half the year's end I'm not where I thought I*
>   *would be."*

Pietry, also called Piphology, and no, Pietry is not a spelling mistake. It is a form of prose where the length of each word corresponds to the digits of Pi, in order. As far as I know, the first piem was written by the British physicist Sir James Hopwood Jeans about a hundred years ago:

> *"How I want a drink, alcoholic of course, after the*
> *heavy chapters involving quantum mechanics!"*

Note the number of letters in each word: 3, 1, 4, 1, 5, 7, 2, 6, 5, 3, 5, 8, 9, 7, 9, the first digits of Pi: 3.141572653589... Jeans wrote the piem to serve as a mnemonic to remember the digits of Pi.

Fibs are poems where the number of syllables in each line goes according to the *Fibonacci series*, where each number is the sum of the two numbers preceding it: 1, 1, 2, 3, 5, 8, 13... ($2 = 1 + 1$, $3 = 2 + 1$, $5 = 3 + 2$, $8 = 5 + 3$, $13 = 8 + 5$, etc.). Books have been written about this amazing sequence, but in this chapter, we only touch on its use as a scaffolding for poetry. Once again, we have Sarah Glaz to thank for the most informative review of Fibs and other mathematically structured literature. Glaz points to screenwriter Greg Pincus as the person who invented Fibs in 2006, but one should also mention Athena Kildegaard, who published a whole book of Fibs the same year that Pincus coined the term. Some great Fibs and lots of great mathematical poetry have been written by another Banff workshop participant, JoAnne Growney, a now-retired math professor who has become a leading math poet, artist, and translator. Here are two Fibs that she kindly allowed me to share with you. In the first poem, "In the Tuscan Sun", the number of syllables in each line is according to the Fibonacci sequence: 1, 1, 2, 3, 5, 8... In the second, "Fibbish!", the number of *letters* in each line increases and then decreases according to the Fibonacci sequence. This poem's increase-then-decrease pattern resembles the growing-melting pattern of "snowballs", also depicted in the visual structure of the poem.

In the Tuscan Sun

"View

the

statue

in Pisa

of Fibonacci,

mathematician in the sun."

Fibbish!

O,

I

go

for

mathy

patterns,

poems

set

on

1,

1.

JoAnne Growney has the most wonderful website dedicated to "poetry, mathematics, art and translation", and her blog "Intersections — poetry with mathematics" is very popular and well worth a visit. You'll find the links in the bibliography.

The people who took the intersections between math and language to extreme limits were the group of French writers and mathematicians known as the Oulipo, short for "Ouvroir de littérature potentielle", which means "workshop of potential literature". Founded in Paris in the 1960s by French novelist Raymond Queneau, the Oulipo proposed novel literary techniques that were mathematically structured using a constraining algorithm. One such method is known as the "N + 7" method. Take a text and replace every noun with the seventh word after it in the English dictionary. The new text is an Oulipian creation exemplifying the group's belief that creativity can flourish, not despite limitations but because of them. Here's the "N + 7" text of the beginning of Jane Austin's *Pride and Prejudice*, generated using Spoonbill, a free online N + 7 generator:

> *"It is a tuber universally acknowledged that a single mandible in postcard of a good founder must be in want of a willingness."*

The original sentence is:

> *"It is a truth universally acknowledged that a single man in possession of a good fortune, must be in want of a wife."*

Georges Pérec was another creative member of the Oulipo. One of his famous works, "Le grand palindrome", is a 1247-word French *palindrome*. Palindromes are words that read the same backward or forward, like radar or "Madam, in Eden, I'm Adam". Numbers can also be palindromes. Mathematicians study number palindromes, on the lookout for interesting phenomena. There is even an *open problem* in math related to palindromes that is still waiting to be solved. The so-called "196 problem" is about a well-known procedure for generating numerical palindromes:

- Add a whole number, say 69, to its reverse 96: $69 + 96 = 165$.

- Do the same with the new number — take it and add it to its reverse: $165 + 561 = 726$.

- Keep going! $726 + 627 = 1353$, $1353 + 3531 = 4884$.

- We get a palindrome!

Does this work with all whole numbers? Do we always get a palindrome? Are there numbers for which we will never get a palindrome? Mathematicians highly suspect that the number 196 is a *Lychrel number* — a decimal[1] number that will never produce a palindrome with this procedure — but haven't yet[2] been able to prove this.

The most recent, and perhaps most mathematical, evolution of doublets is Wordle™. Wordle™ was invented by software engineer Josh Wardle in 2013 and further developed by his partner Palak Shah in 2021. It instantly became popular and was acquired by *The New York Times* in 2022. You'll find many online Wordle™ games to play. It is a one or two-player game played on an originally empty 5-column, 6-row grid, probably best described as a cross between Doublets and Mastermind™. The player's goal is to guess a 5-letter word within six tries. A word is chosen either by her opponent or a computer (for the one-player version) and concealed from the first player. She makes her first guess and writes the word in the top row.

---

[1] I'm carefully referring to the decimal (base-10) system we are familiar with. Lychrel numbers do indeed exist in other number base systems. One example is the binary number: 10110. It has been proven that this never generates a palindrome using the "add to reverse" procedure.

[2] To date: July 13, 2023

The computer, or opponent, scores her first guess according to the following rules:

- Letters that appear in the concealed word in the correct position are highlighted green.

- Letters that appear in the concealed word but are not in the correct position are highlighted yellow.

- Letters that do not appear in the concealed word are highlighted grey.

The player then adjusts her guess accordingly and writes it in the second row. Her opponent scores the guess, and this routine continues until the table runs out of rows or until the player guesses the word, in which case she wins the game.

Figure 2.2 shows a Wordle™ game where the player guessed the hidden word after four tries.

Fig. 2.2 A solved Wordle™ puzzle

The scoring system reminds us of Mastermind™, while the rearranging of the correct letters in each new guess and the restricted number of guesses resembles Doublets. Math helps us answer many questions we might ask about Wordle™. Given that in English, at least in most versions, there are 2,315 possible 5-letter solutions, it might seem impossible to guarantee a win in any given puzzle. However, Dimitris Bertsimas and Alex Paskov from MIT have found an algorithm that does just this, guaranteeing a win within the 6 allotted guesses. They found that by using the word SALET, which means helmet, as the first word, they can create what's known as a decision tree that gives directions as to what the next guesses should be in every step according to all possible scores. For example, their algorithm suggests that if the score after SALET is five greys, meaning none of the letters are correct, the next guess should be COURD. If the response to this is a yellow U and four greys, meaning that out of the five letters COURD, only the U appears in the hidden word — and it is not in the middle — the next guess should be FUNNY. If the response to this is greens for F, U, and Y, meaning that these letters are correct and in the right place, then the hidden word *must* be FUZZY because this sequel of scores

$$5 \text{ greys} \rightarrow \text{yellow U} \rightarrow \text{green F, U and Y}$$

is the only one that leads from SALET to FUZZY. Other researchers have also published similar strategies. All agree that all traditional Wordle™ English puzzles (those with 2,315 possible solutions) can be solved within the 6-row constraint, yet it remains to be proven[3] or disproved if they can be solved with less, perhaps by using a different first word.

Decision trees, such as the one used in Bertsimas and Paskov's algorithm, are often used in math and computer science to help — well — make better decisions. As long as a problem at hand and the possible paths to its solution can be written down as a mathematical object, they can be employed to help find the best solutions. The mathematical object used is known as a graph, and the theoretical

---

[3] To date: July 13, 2023

framework behind it is called graph theory. The graphs in graph theory are not to be confused with the graphs we are all familiar with, like the *x-y-z* Cartesian coordinate system, pie charts, etc. They are best described as a diagram made out of points known as vertices and lines connecting them, known as edges. The vertices represent data, and the edges represent relationships. In Bertsimas and Paskov's decision tree, each vertex represents a guess from which edges emanate according to all possible scores, with one edge for each score. Each edge leads to another vertex, which is the next guess. In effect, Bertsimas and Paskov found that the *depth* of the tree — the number of levels needed to reach all the 2,315 final solutions = vertices — when the first guess is SALET, is 6.

Figure 2.3 shows a schematic sketch of Bertsimas and Paskov's decision tree.

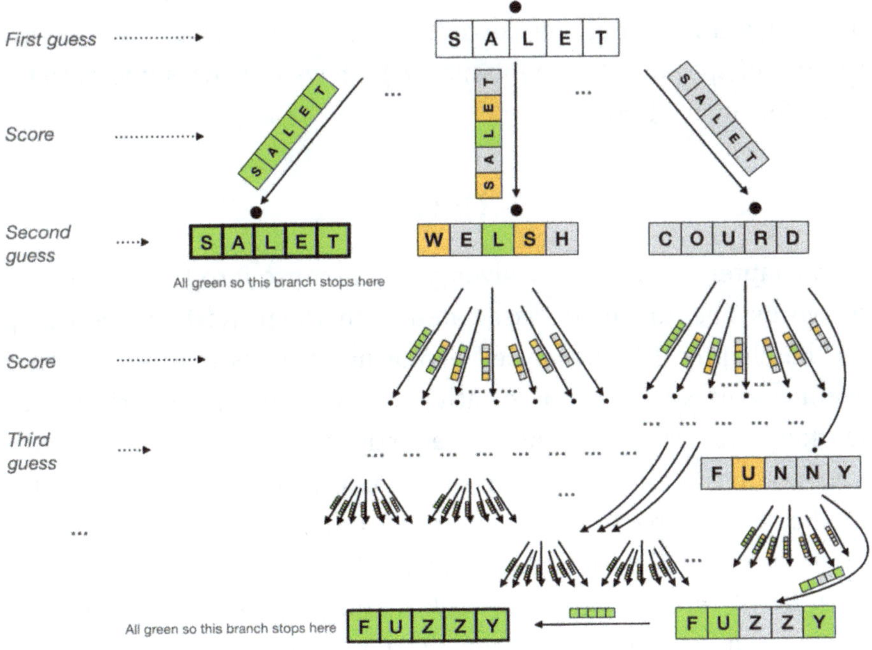

Fig. 2.3　A sketch of Bertsimas and Paskov's decision tree

It's perhaps important to note that graphs are not used just for making decisions. They are extremely useful tools in many contexts, including other important literary contexts besides puzzles. They are used, for example, to create and design the interactions between different characters in a story. Authors, linguistic editors, and grammar-correcting software also use them in many cases to work out the correct syntactic structure for any given sentence.

Many modifications can be made to doublets, Wordle, and other similar puzzles. For doublets, we can change the constraints by allowing the player to change more than one letter in each step, change a letter and its position in each step, or even alternate: change a letter in one step and its position in the next. Wordle™ can be modified to words with four, six, or seven letters, or we can allow the opponent to change the concealed word according to the player's responses, as long as the scores along the way are still correct, a version known as Absurdle. We can add a score for letters one place away from the correct letter. Unfortunately, we've well run out of space in this chapter, so we'll have to leave some of these ideas for my next book.

Recap

This chapter has hopefully given you a glimpse into the wild world of words and the math associated with them. As I was working on the chapter, I found myself exploring avenues that led to other avenues and paths... my associative way of thinking and writing was working overload! I have not done justice to this topic that deserves a whole series of books. However, going from doublets to poetry to the philosophy of language, and from a nearly 4,000-year-old math poem to pictorial math, piems, Fibs, and the Oulipo, looping back to Wordle™ at the end, took us on perhaps a rather eclectic and hopefully interesting tour of the topic. Along the way, we encountered some important concepts:

- *Doublets* — also known as *Word Ladders*, a game invented by Lewis Carroll, where you have to transform one word into another with the same number of letters in as few steps as possible by replacing one letter in each step without changing the order of the letters.

- *Anagrams* — given a text of any length, its anagram is a text of the same length with the same letters re-arranged to form a new text.

- *Recreational linguistics* — exploring language through wordplay, puzzles, and games.

- *Language* — subjective and culturally formed complex mental constructions we use to organise and understand our experiences. Language expresses concepts by turning them into discrete symbols, words, or phrases that can be communicated to others.

- *Haiku* — a 3-line poem form where the first and last lines have five syllables, and the middle line has seven syllables.

- *Pietry or Piems* — poems where the number of letters in each word is equal to the corresponding digit in the decimal form of Pi: 3, 1, 4, 1, 5, 9...

- *Fibs* — poems where the number of syllables in each line goes according to the numbers in the *Fibonacci series* in order.

- *Oulipo* — a group of French writers and mathematicians known for their algorithmic construction of literature.

- *N + 7* — a literature-generating algorithm employed by Raymond Queneau, founder of the Oulipo, in which each noun in a given text is replaced by the noun that is seven places further on in a given dictionary.

- *Palindromes* — any text or number that reads the same forward and backward.

- *Open problem* — a problem that is as yet unsolved.

- *Lychrel number* — a decimal number that will never produce a palindrome using the iterative procedure of repeatedly adding a number with a digit-reversed duplicate copy of itself.

- *Wordle*™ — a word puzzle where a player has to guess a concealed 5-letter word within six guesses using the information provided by his opponent (or a computer) after each guess. The information is a colour code for each of the guessed letters. Those correct and in the correct position are coloured green, those correct but not in the correct position are coloured yellow, and those not in the hidden word are coloured grey.

- *Absurdle* — a version of *Wordle*™ where the concealed word can be changed by the opponent on his turn as long as the new word still abides by all the scores given from the beginning of the game.

- *Graph theory* — an area of math used to analyse items that have some form of relationship between them.

- *Decision tree* — a mathematical graph that depicts situations that branch into different outcomes according to a logical test applied to each situation.

Challenge Yourself!

(1) Solve the following doublets. In some cases, there is no way to verify that the number of steps is minimal — so if you find a solution with fewer steps than I did, please let me know!
- Turn SICK into WELL in at most four steps.
- Turn EASY into HARD in at most four steps.
- Turn NERD into COOL in at most seven steps.

(2) There are at least 12 anagrams for the word THREAD. One of them is HE DRAT. Find the other 11. Note that rearranging the order of words is not counted, so DRAT HE would not be one of the 12.

(3) Following are some anagrams of some well-known 3-word sentences. Find the original sentences.

- PA MIGHT END
- A WEE SYNC
- READERS COMMUTE
- THE TITANS CARD SONG

(4) Using an online translator, the following nouns have been translated from English into Hebrew, into Swahili, into Japanese, and then back into English. Find the original word without doing the reverse process, although even trying to the reverse the process in many cases won't work (why?)... and good luck!

- Mirror
- Seaweed
- Rack
- Ball

(5) A Wordle™ player tried the following guess: JECUS despite the fact that it isn't a recognised English word. To his surprise, he got the following score: J → Yellow, E → Grey, C → Yellow, U → Green, S → Grey. To his even greater surprise, he was able to guess the correct word in his next guess. What was the word?

Solutions

(1) • SICK → SILK → SILL → WILL → WELL.

   • EASY → EAST → HAST → HART → HARD.

   • NERD → HERD → HELD → HELL → TELL → TOLL → TOOL → COOL.

(2) Anagrams of THREAD:
   • Dearth
   • Thread
   • Hatred
   • Drat Eh
   • Dart Eh
   • Hart Ed
   • Tad Her
   • Red Hat
   • The Rad
   • At Herd
   • He Dart
   • He Drat

(3) The original sentences:
   • PA MIGHT END → MIND THE GAP
   • A WEE SYNC → YES WE CAN
   • READERS COMMUTE → DREAMS COME TRUE
   • THE TITANS CARD SONG → NO STRINGS ATTACHED

(4) • Mirror — originally: Glass
   • Seaweed — originally: Glue
   • Rack — originally: Fortune
   • Ball — originally: Pill

(5) The word was CAJUN

Bibliography and Further Reading

Anand, M. (2015). *A New Index for Predicting Catastrophes*. McClelland & Stewart.

Babin-Fenske, J., Anand, M., and Alarie, Y. (2008). Rapid morphological change in stream beetle museum specimens correlates with climate change. *Ecological Entomology* **33**, 5, 646–51.

Bertsimas, D. and Paskov, A. (2022). An exact and interpretable solution to Wordle. Retrieved July 21, 2023, https://auction-upload-files.s3.amazonaws.com/Wordle_Paper_Final.pdf

Carroll, L. (2006). *Alice in Wonderland*. Urbana, Illinois: Project Gutenberg. Retrieved July 21, 2023, www.gutenberg.org/ebooks/19033.

Chapman, R. (1988). *Distance, Rate, Time*. Fireweed Press. USA.

Gardner, M. (1977). Mathematical games. *Scientific American*, **236**, 2, 121–27. http://www.jstor.org/stable/24953898

Gardner, M. (1996). Word ladders: Lewis Carroll's doublets. *The Mathematical Gazette*, **80**, 487, 195–98. https://doi.org/10.2307/3620349

Glaz, S. (2016). Poems structured by integer sequences. *Journal of Mathematics and the Arts*, **10**, 1–4, 44–52.

Growney, J. (2023a). Intersections — Poetry with mathematics. *Blogspot*. Retrieved August 1, 2023, https://poetrywithmathematics.blogspot.com/

Growney, J. (2023b). Poetry, math, art, translation. Retrieved August 1, 2023, https://joannegrowney.com/

Padgett, R. (2000). *The Teachers & Writers Handbook of Poetic Forms*. 2nd edition. T & W Books. New York.

Pinker S. (2007). *The Stuff of Thought: Language as a Window into Human Nature*. Viking.

Watts, I. (2011). Against idleness and mischief. In: *Watt's Songs Against Faults*. Urbana, Illinois: Project Gutenberg, pp. 18–19. Retrieved July 21, 2023, https://www.gutenberg.org/ebooks/37543.

Weisstein, E. W. (2002). Lychrel number. *From MathWorld — A Wolfram Web Resource.* Retrieved August 1, 2023, https://mathworld.wolfram.com/LychrelNumber.html.

Wikipedia contributors (2023). Piphilology. *Wikipedia, The Free Encyclopedia.* Retrieved August 4, 2023, https://en.wikipedia.org/w/index.php?title=Piphilology&oldid=1144629006.

Wikipedia contributors (2023). Wordle. *Wikipedia, The Free Encyclopedia.* Retrieved August 4, 2023, https://en.wikipedia.org/w/index.php?title=Wordle&oldid=1168219177.

Chapter 3

Speed Math

Introduction

One of the best ways to amaze and amuse your friends is to perform seemingly impossible mathematical feats in your head. Why not try some out yourselves? No cheating!

The Puzzle

Calculate the following expressions in your head. No calculators!

- *The sum of all the numbers between 1 and 100 (included)*

- $35^2 = ?$

- $92^2 = ?$

- $12\% \times 250 = ?$

Where to Start?

If you try to calculate these in the usual way, you'll probably get stuck. On the other hand, if you already know some of the tricks, it'll be too easy. If you really want to try and solve these on your own and without prior knowledge, you will have to spend some time pondering about the problems instead of just running in and calculating. Here are two key concepts that might help. The first is symmetry. Try and see if you can reconstruct the problem in a different way than presented. This is useful for the first and last calculations. The second is to work out the algebra of the general problem and see if this leads you anywhere. This is useful for the second and third problems.

Solving the Puzzle

- *The sum of all the numbers between 1 and 100 (included)*

  Let's backtrack a little and solve an easier problem first: calculate the sum of all the numbers between 1 and 10 (included). We can do this sequentially: $1 + 2 + 3 + 4 + 5 + 6 + 7 + 8 + 9 + 10$, with the sum accumulating in our head accordingly: 1, 3, 6, 10, 15, 21, 28, 36, 45, 55. There is, however, an easier way that we show pictorially in Fig. 3.1. Pairing the numbers from the outside in, we see a nice symmetry: five pairs of numbers that all sum up to 11: $1 + 10, 2 + 9, 3 + 8, 4 + 7, 5 + 6$.

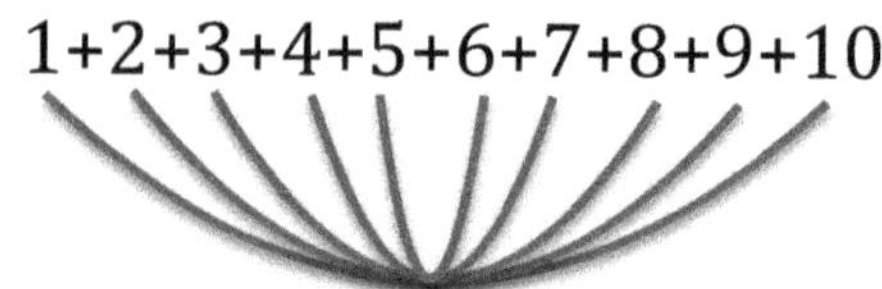

Fig. 3.1   Gauss' method for speed addition of the numbers 1 to 10

This method is known as Gauss' method. First, instead of calculating $1 + 2 + 3 + 4 + 5 ... + 97 + 98 + 99 + 100$, mentally group the numbers into the 50 pairs, each of which equals 101:

$$(1 + 100), (2 + 99), (3 + 98), (4 + 97), (5 + 96)...(50 + 51)$$

The sum is, therefore,

$$50 \times (100 + 1) = 50 \times 101 = 5050 \tag{3.1}$$

In general, the sum of a series of consecutive whole numbers equals half the length of the series times the sum of the first and last numbers. So, the sum of the series of length 8: 1, 2, 3, 4, 5, 6, 7, 8, is equal to:

$$4 \times (1 + 8) = 4 \times 9 = 36 \tag{3.2}$$

and the sum of the series of length 9: 1, 2, 3, 4, 5, 6, 7, 8, 9, is equal to:

$$4\frac{1}{2} \times (1 + 9) = 4\frac{1}{2} \times 10 = 45 \tag{3.3}$$

For series of odd lengths (like 9), it's usually easier mentally to multiply the sum of the first and last numbers by half "the length of the series minus 1" and then add the middle number. Instead of $4\frac{1}{2} \times (1 + 9) = 45$, use:

$$\frac{1}{2} \times (9 - 1) \times (1 + 9) + 5 = \frac{1}{2} \times 8 \times 10 + 5 = 4 \times 10 + 5 = 45 \tag{3.4}$$

- $35^2 = ?$

The easiest way to do this mentally is to use the following recipe, which works for squaring any number that ends in a 5:

- The first digits of the answer will be the digit — or digits for numbers with more than two digits — multiplied by its consecutive whole number.

- The last digits of the answer will be 25,

and that's it! Let's see how this works with the number 35:

- The first digits of the answer will be $3 \times 4 = 12$.

- The last digits are: 25
    $\longrightarrow$ The answer is: 1,225

Notice that since 3 is in the tens column, when you multiply it by the whole number that follows, 4, you are really multiplying 30 by 40 to give 1,200 — a number that divides 100 without remainder. Generalising this for other numbers that end in 5, we can understand why the last digits are *always* 25; the multiplication of the first digit by its consecutive whole number is a multiple of 100, so they can't change the 25 in the tens or units column. Let's see how the recipe works for some other numbers.

Here's how it works with the number 75:

- The first digits of the answer will be $7 \times 8 = 56$.

- The last digits are: 25
  $\rightarrow$ The answer is: 5,625

And here's how it works with the number 115:

- The first digits of the answer will be $11 \times 12 = 132$.

- The last digits are: 25
  $\rightarrow$ The answer is: 13,225

Squaring 5s this way is one of the neatest tricks of the trade. Why does it work? There are a few ways to analyse the calculation, but I like the *pictorial* one shown in Fig. 3.2. The area of the square is 35×35, and it can be rearranged into a 30×40 rectangle and a separate 5×5 square. Summing the areas, we get $1200 + 25 = 1225$. Note also that $1200 = 3 \times 4 \times 100$ — 100 times the product of 3, and its consecutive whole number, 4. Although the square and rectangle in Fig. 3.2 are specific for the special case of 35^2, they can easily represent the general case if we remove the labels since squares of any size can always be transformed in this manner, as long as their length is a multiple of 5.

We can also use algebra to show how this method works. We'll start with the original challenge: $35^2 = ?$, but we'll separate the

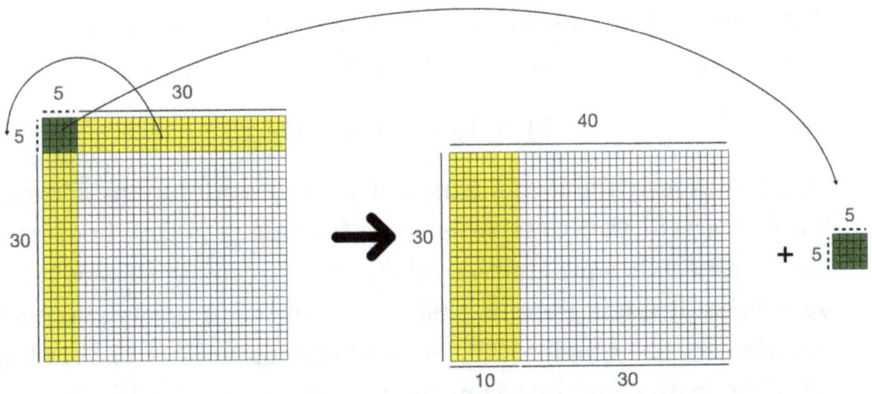

Fig. 3.2   Pictorial explanation why $35^2 = 30 \times 40 + 25$

tens and the units digits and write it as a multiplication problem like this:

$$(30 + 5) \times (30 + 5) = ? \tag{3.5}$$

Now, we open the brackets to get:

$$(30 + 5) \times (30 + 5) = 30 \times 30 + 2 \times 30 \times 5 + 25 \tag{3.6}$$

We'll rearrange the numbers back into brackets in a few steps, leaving out the number 25:

$$\begin{aligned}
(30 + 5) \times (30 + 5) = 30 \times 30 + 2 \times 30 \times 5 + 25 &= \\
30 \times (30 + 2 \times 5) + 25 &= \\
30 \times (30 + 10) + 25 &= \\
100 \times 3 \times (3 + 1) + 25 &= 1225
\end{aligned} \tag{3.7}$$

Take a minute to ponder on the math. First, at least for the number 35 in the example, we can see that the answer's last digits are 25, and the first digits are 3 times "its consecutive number", 4. Now, we can generalise. Any number ending in 5 can be written as a sum of a multiple of ten and the number 5. By denoting the digits multiplied by 10 with the letter x so that the number squared is $(10x + 5)$, we will always get the same format:

$$(10x + 5)^2 = 100 \times x \times (x + 1) + 25 \qquad (3.8)$$

which is exactly the formula used. This will work for any x, even if it's a fraction. Of course, when it *is* a fraction or when x becomes larger, it becomes prohibitively difficult to calculate $x \times (x + 1)$ in your head. It might then be easier to open the brackets and just calculate $x \times x + x$, so for 125^2, we would calculate $12^2 + 12 = 144 + 12 = 156$ instead of 12×13 and get the answer: $125^2 = 15625$.

- $92^2 = ?$

  Here's a nice trick for numbers close to 100:

  - First, note that for numbers larger than 31, the result will always have four digits. Since this method is only practical (at least mentally) for numbers close to 100, we'll assume that the result has four digits.

  - The first two digits: subtract from the given number the difference between the number and 100. For 92, this difference is $100 - 92 = 8$, so subtract 8 from 92: $92 - 8 = 84$.

  - The last two digits are the difference between the given number and 100, squared: $8^2 = 64$.
    $\longrightarrow$ The answer is 8,464.

  Here's how it works with the number 97:

  - $100 - 97 = 3$. The first digits of the answer will be $97 - 3 = 94$.

  - The last digits are $3^2 = 9$.
    $\longrightarrow$ The answer is 9,409 (note the added 0 in the tens column — the result has four digits).

  And here's how it works with the number 89:

  - $100 - 89 = 11$. The first digits of the answer will be $89 - 11 = 78$.

- The last digits are $11^2 = 121$. Oops! 121 has 3 digits! In cases like this, the last digits are the last two digits (21), and the first digit (1) is "carried" and added to the first two digits. So the first two digits are $78 + 1 = 79$, and the last two are 21.

  $\longrightarrow$ The answer is 7,921.

We can justify this pictorially, as in Fig. 3.3, or by using algebra.

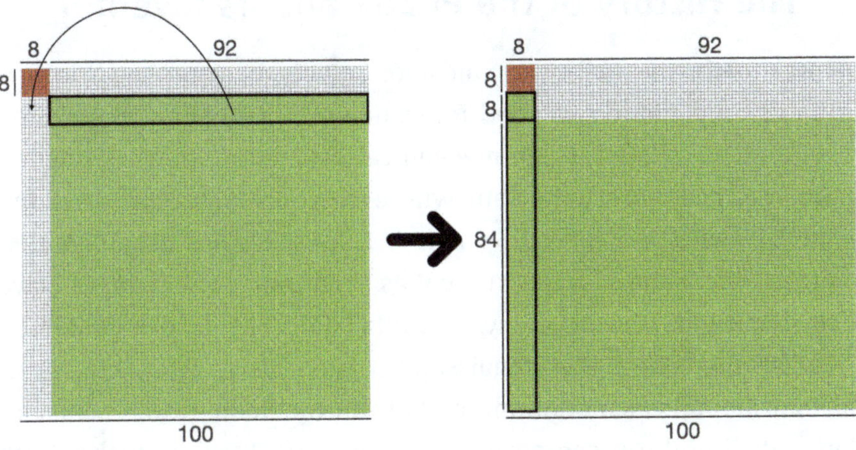

Fig. 3.3 Pictorial explanation on why $92^2 = 84 \times 100 + 8 \times 8$. The red 8×8 square in the top left corner signals the difference from 100

Rewriting 92 as $100 - 8$, we ask: $(100 - 8) \times (100 - 8) = ?$

Opening the brackets, we explicitly get the formula that we used:

$$(100 - 8) \times (100 - 8) = 10000 - 8 \times 100 - 8 \times 100 + 64 = 8464 \qquad (3.9)$$

- $12\% \times 250 = ?$

The trick here is to rewrite the expression explicitly as multiplication and use the fact that the order of the multiplicands doesn't matter:

$$12\% \times 25 = \frac{12 \times 25}{100} = \frac{25 \times 12}{100} = 25\% \times 12 = 3 \qquad (3.10)$$

When multiplying two numbers, especially with percentages, look for the easiest way to calculate. Although it might sound counterintuitive, the math is sound! 12% of 25 is just the same as 25% of 12, that is, a quarter of 12, equalling 3.

The History of the Puzzle and its Inventor

Mental math, the ability to calculate rapidly in your head, is fun, fulfilling, and, some say, good for your brain. Of course, before the invention of calculators or mechanical calculating devices, mental math was also very important wherever you didn't have or didn't want to use some form of writing. It is therefore assumed that "human calculators", also known as "lightning calculators", have always existed. Unfortunately, I couldn't find any record of mental calculations before the Renaissance. There have been claims of ancient mental calculation methods from the Vedic texts — ancient Sanskrit literature composed between the 14th and the 10th centuries B.C.E. These Vedic techniques appeared in a book called *Vedic Mathematics*, written during the 20th century by Bharati Krishna Tirtha, an Indian monk who also held a Masters degree in mathematics. Although some of the tricks in the book are quite impressive, others are well-known and similar to European techniques from the last few hundred years. Many prominent scholars have debunked Krishna Tirtha's association of these techniques with ancient Vedic manuscripts.

We know a lot about the more recent history of human calculators through the extensive work of British mathematician Walter Rouse-Ball. In the late 19th century, Rouse-Ball published one of the first exclusive recreational math books. Titled *Mathematical Recreations and Essays* and updated at the beginning of the 20th century by the British-Canadian mathematician Donald Coxeter, it quickly became a classic text, and to date, 13 editions have been published. Chapter 13 focuses on calculating prodigies, young kids who were able to do remarkable mental calculations, like multiplying, squaring, and dividing very large numbers or solving real-world problems. Most of these kids showed their talent at a very young age, even before they learned to read or write, with the effect waning as they grew older. We're unsure how they did this; however, they all seem to have had an extraordinary memory. Some of them may have had the "Rain-man" (Savant) syndrome or something similar.

Renaissance human calculators thrived throughout the 17th and 18th centuries. We've already seen the speed addition method devised by German mathematician and physicist Johann Carl Friedrich Gauss (Fig. 3.4). Born in 1777 to a father who held all kinds of odd jobs, among them gardener and bricklayer, and an illiterate mother, Johann Gauss was a child prodigy. In fact, the first puzzle I

Fig. 3.4   Carl Friedrich Gauss

gave you was a real classroom exercise that was given to Gauss and his classmates at school by his teacher — and Gauss amazed both the teachers and his classmates by calculating the correct answer in his head within seconds. Well, at least that's one version of the story based on one of his many biographies. Gauss became one of Germany's greatest mathematicians and contributed much to the development of algebra, number theory, and numerical analysis, as well as differential geometry, astronomy, magnetism, and a whole bunch of other stuff. He would have published much more, except that he was such a perfectionist that he rarely published his work. Gauss preferred to derive his work independently instead of relying on work done by other mathematicians. He was meticulous about every single word that he wrote and about giving correct credit to the mathematicians upon whose ideas he built. So much so that, in many cases, he preferred not to give credit at all — to the annoyance of his fellow mathematicians — because it took so long to trace the exact history of the idea. His perfectionism, his obsession with exploring and constructing mathematical tables, and his hobby of collecting number trivia well reflect his early years as a math prodigy. Gauss became a celebrated mathematician at the University of Göttingen and was well-known for his intense research. He hated lecturing and didn't care much for popularising his work, so he probably wouldn't approve of this book, especially since it perpetuates his childhood speed addition method (that probably wasn't even his original idea) rather than discuss his many serious contributions to math. Gauss lived through many personal tragedies, including the death of his two wives, a son and a daughter, and a continuing conflict with his remaining sons. He died of a heart attack in Göttingen in 1855.

During the Victorian era, especially in England and the United States (US), many child prodigies gained quite a lot of fame, mostly because their parents took advantage of them to make quick money by showing off their extraordinary feats to the public. In at least one case, two prodigies, 14-year-old American-born Zerah Colburn and

12-year-old British-born George Parker Bidder, were pitted against each other at a calculating duel in London in 1818 (Bidder won). Bidder and Colburn are better known than other human calculators of their time, mainly because they both explained their methods, giving insight into the math behind them and paving the way for others. Mental calculators and magicians use their methods to this day.

Ball and Coxeter mainly discuss child prodigies; however, many professional mathematicians were and are very capable mental calculators. Some, like Bidder, Irish-born William Rowan Hamilton and New Zealand-born Alexander Aitken, started off as child prodigies, retained their talent as they grew up, and went on to become prominent engineers, mathematicians, and scientists. Others, like the 17th-century Oxford math professor John Wallis, developed them later on in life. Wallis, one of the fathers of integral calculus, correctly mentally calculated the square root of a 53-digit number one night when he had trouble sleeping! The 20th-century polymath and nuclear physicist John Von Neumann was also a lifelong prodigy and had an extraordinary memory. As a young child, he could instantly divide two 8-digit numbers in his head, and later on, he could recall extensive passages from books after reading them just once!

Some of the human calculators were able to produce some very elaborate math. Indian mathematician Srinivasa Ramanujan (Fig. 3.5) was one of the most famous of these prodigies. Born in 1877, he grew up in Kumbakonam, a small town in India. Ramanujan had no formal training in math, yet he went on to become one of India's, if not the world's most famous, mathematicians, making important discoveries in many areas of math. His short life — he died at the age of 32 from complications of tuberculosis (or possibly misdiagnosed amoebiasis) — had many ups and downs. He excelled in math from a very young age and flourished throughout his school years by studying college-level textbooks and winning various

Fig. 3.5   Srinivasa Ramanujan

local prizes along the way, yet failed miserably in securing a formal degree because of his disinterest in anything but math. His health and wealth as a young adult were frequently in jeopardy, but despite this, he managed to pour a voluminous number of mathematical formulae into notebooks, many of which were unknown, and practically all of them derived from scratch. His mathematical fame came when Indian mathematicians who recognised Ramanujan's brilliance sent his notebooks to British mathematicians, including Godfrey Harold Hardy, who later became Ramanujan's mentor and collaborator. It took Hardy some time to decipher the notebooks and recognise the mathematical ingenuity of their author, particularly because many discoveries seemed to be intuitively thought out and lacked rigorous proof. However, once he did, he persuaded Ramanujan to work with him at Cambridge, where he stayed until his untimely death. Hardy, himself one of Britain's prominent mathematicians, when asked what his greatest achievement was, remarked that it was "discovering Ramanujan!"

Ramanujan's life, his association with Hardy, and his extra-ordinary mathematical talent have been celebrated in books, films, and popular culture. The 2015 film *The Man Who Knew Infinity*, based on Robert Kanigel's novel of the same name, was lauded by mathematicians as doing justice to both mathematicians and their mathematical accomplishments.

Ramanujan's phenomenal calculations differed from those of Gauss, Bidder and Aitken, in that they related to unusual characteristics of mathematical entities. He was able to instantly decompose large numbers into their prime factors, like $538 = 2 \times 269$, and came up with a highly accurate approximation for pi. A well-known example of his seemingly magic mental math is the Hardy–Ramanujan "taxicab" number 1729. When Hardy visited Ramanujan in Putney, he mentioned that the taxi's number was 1729, a number that did not seem to be of any mathematical significance. Ramanujan reportedly replied, "It is a very interesting number; it is the smallest number expressible as the sum of two cubes in two different ways." And indeed,

$$1729 = 9^3 + 10^3 = 1^3 + 12^3 \tag{3.11}$$

One of the more inspiring stories of adult mental calculators is that of Jakow Trachtenberg (Fig. 3.6). Trachtenberg was not a child prodigy. Born in Odessa, Russia, in 1888, he trained as an engineer and worked first in St. Petersburg and then in an arms factory in Obukhov, quickly becoming its chief engineer. But Trachtenberg also quickly became an outspoken pacifist, and it was this that got him into trouble, first with Communist Russia and then with Nazi Germany. For much of his adult life, his biography is full of narrow escapes from the clutches of evil regimes. He fled from Russia to Germany to escape the communists and from Germany to Austria to escape the Nazis until they caught up with him and threw him into prison. Undeterred and aided by his dedicated wife, Alice,

Fig. 3.6   Jakow Trachtenberg photographed on April 23, 1940, by the Vienna Gestapo

while continuously advocating for world peace, he managed to escape prison and live as a fugitive in Yugoslavia, only to be caught in the late thirties by the Nazis and sent as a political prisoner to a concentration camp. Trachtenberg spent 7 years in the camp, enduring forced labour and living under the most inhumane conditions imaginable. It was during these years that Trachtenberg developed some special methods for speed calculations of basic math, using only his mind — since there was, of course, no pencils or paper in the camp.

In May 1944, after learning that her husband was about to be executed, Alice Trachtenberg arranged for Jakow's transfer from the camp to another camp in Leipzig by bribing senior officials. Trachtenberg managed to escape this camp by crawling through double barbed-wire fences, only to be caught again and sent to yet another forced labour camp in Trieste, Italy. Trachtenberg managed to escape from this camp as well, aided once more by his dedicated wife, and together, they finally made it to freedom, fleeing to Switzerland. After the war, Trachtenberg dedicated himself to education, founding the Mathematical Institute in Zurich, where he perfected his method and taught it to children and adults alike. Trachtenberg's speed system of basic mathematics quickly became popular, and Jakow Trachtenberg became convinced that by mastering it, anyone could become a human calculator. The system is unique because it is one complete system that can be used in all basic arithmetic calculations. It is also very simple, quick, and easy to learn. If you are interested in this system, you can learn it from the book *The Trachtenberg Speed System of Basic Mathematics*. To give you some idea of his system, here is a short description of his method for multiplication by 11 and 12.

Multiplying a long number by 11:

- Add a leading zero to the number

- Working from right to left, write underneath each digit its sum with its neighbour to the right. The units digit of the number has no neighbour to the right, so you "add zero" and write it down as the first digit of the answer. If the sum is larger than 9, carry the tens digit and add it to the next digit you write down.

Example: Multiply 9432423 by 11:

- $09432423 \times 11 =$

- 103756653

Multiplying a long number by 12:

- Add a leading zero to the number

- Working from right to left, write underneath each digit double the digit plus its neighbour to the right. The units digit of the number has no neighbour to the right, so you "add zero" and write it down as the first digit of the answer. If the sum is larger than 9, carry the tens digit and add it to the next digit you write down.

Example: Multiply 9432423 by 12:

- $09432423 \times 12 =$

- 113189076

Once you get the hang of it, you'll find you don't need to write down the answer — you can just run off the digits in your head, from right to left.

The last person I want to mention in this section is Aaryan Nitin Shukla (Fig. 3.7). Not that he's made waves as a great mathematician, which would be — well — rather presumptious since he's only 13 years old. He has made waves, though, as a child prodigy, being the youngest ever mental maths calculator to win the Mental Calculation World Cup in 2022. The World Cup is run by the Global Mental Calculation Association, which arranges international speed addition, multiplication, division, squaring, and speed calendar calculation competitions. Born in Nashik, India, Shukla has been doing mental math since he was six. He has won many prizes, frequently appears in the media, and aspires to be either an engineer or a mathematician when he grows up.

Fig. 3.7   Aaryan Nitin Shukla

Generalisations and Further Explorations

There seems to be no end to speed math methods. Every now and again, some new techniques keep on popping up on social media.

US mathematician Arthur Benjamin is famous for inventing and performing speed math. He once taught me this method for mentally squaring the number 479.

- Look for the nearest hundred — for 479, this is 500.

- Note the positive difference between the number and its nearest hundred. $500 - 479 = 21$.

- Subtract this difference from the given number. $479 - 21 = 458$.

- Multiply the result by the nearest hundred. $458 \times 500 = 229000$.

- Add the square of the difference you found earlier. $229000 + 21^2 = 229000 + 441 = 229441$.

- That's the answer! $479^2 = 229441$.

Here are a few comments about the method. First, you need to find the nearest hundred because multiplying by 100 is an easy task. Second, when looking for the nearest multiple of 100 you can either go up or down; for example, if you want to square 118, you should go down to 100, not up to 200 (though both will work). If you go down, then in the next step, you add the difference to the given number instead of subtracting it (so for 118, you add 18 to 118 to get 136). Benjamin's method is general and can be used to square any number of any length. except that for squaring a 2-digit number, you look for the nearest multiple of 10; for a 4-digit number, you need the nearest multiple of 1,000, etc.

Sometimes, squaring of the difference is not mentally easy, especially if you're squaring numbers 3-digits long or more. In these cases, just use the technique again to square the difference; for example, if you want to square 837, you'll need to calculate 800×874 mentally and then square 37. But mentally squaring 37 is difficult in itself, so we sidestep and use the method again to first calculate that using $37^2 = 40 \times 34 + 3^2 = 1369$.

Another method that Benjamin employs is for multiplication. Instead of multiplying from right to left, as we are all taught, he does left-to-right multiplication. This gets rid of the need to remember digits that are carried. For simplification, we'll briefly explain how to do this in the case of a many-digit number, like 837, multiplied by a single-digit number. For mental squaring using Benjamin's method, this is always the case since one of the numbers multiplied will always be a single digit followed by zeroes. Multiplying from left to right is performed by breaking up the long number into its power-of-ten constituents, multiplying each constituent by the single-digit number, and adding its sum to the total in order. With a little practice, you should be able to master it! Here is an example of how left-to-right multiplication works to calculate 874×8:

- $800 \times 8 = 6400$

- $+(70 \times 8 = 560) = 6960$

- $+(4 \times 8 = 32) = 6992$

Combining the last two examples, we mentally calculate 837^2 in the following manner:

- Look for the nearest hundred — for 837, this is 800.

- Note the positive difference between the number and its nearest hundred. $837 - 800 = 37$.

- Add this difference to the given number. $837 + 37 = 874$.

- Multiply the result by the nearest hundred. 874×800. Use the left-to-right multiplication method for 874×8 and append 00 as the last digits of the product →

  - $800 \times 8 = 6400$

  - $+(70 \times 8 = 560) = 6960$

  - $+(4 \times 8 = 32) = 6992$
    → $874 \times 800 = 699200$

- Add the square of the difference you found earlier, using the speed squaring method again to calculate 37^2.

- Look for the nearest ten — for 37, this is 40.

- Note the positive difference between the number and its nearest ten. $40 - 37 = 3$.

- Subtract this difference from the given number. $37 - 3 = 34$.

- Multiply the result by the nearest ten. $34 \times 40 = 1360$.

- Add the square of the difference you found earlier. $1360 + 3^2 = 1360 + 9 = 1369$.

  $\longrightarrow 699200 + 37^2 = 699200 + 1369 = 700569$.

- That's the answer! $837^2 = 700569$.

Admittedly, this looks like a long and tedious procedure, but that's only because I've painstakingly written down all the steps. If I had written down only the equations, you'd see it doesn't take up much space. With practice, I assure you, this can be done mentally! Here, for example, is how to employ the method to calculate 118^2 but writing only the appropriate equations:

$$118^2 = 100 \times 136 + 18^2 = 100 \times 136 + (20 \times 16 + 4) = 13924.$$

Arthur Benjamin used one more trick that made squaring 479 much simpler. He used the insight that multiplying by 500 is the same as multiplying by 1,000 and dividing by 2. So a really easy way to calculate 458×500 is $\frac{1}{2} \times 458 \times 1000 = 229000$. This is a great shortcut for numbers that are close to 500. Of course, if the number is close to 1,000, the calculation is even easier. Similarly, you can also use $250 = \frac{1}{4} \times 1000$ for numbers in that vicinity.

Some speed math techniques can be designed based on the following algebraic expressions, where a and b represent two numbers.

- $(a + b)^2 = a^2 + b^2 + 2 \times a \times b$

- $(a + b) \times (a - b) = a^2 - b^2$

Using the first expression, you can break up a number you want to square into two addends that make it easier for you to calculate. For example, by breaking up 308 into $300 + 8$, you'll find that $308^2 = (300 + 8)^2 = 300^2 + 8^2 + 2 \times 300 \times 8 = 90000 + 64 + 4800 = 94864$.

If you want to calculate 13×7, it might be quicker to use the second expression by noticing that $13 = 10 + 3$ and $7 = 10 - 3$, so $13 \times 7 = 10^2 - 3^2 = 91$. Conversely, if you want to calculate $67^2 - 33^2$, you realise that $67^2 - 33^2 = (67 + 33) \times (67 - 33) = 100 \times 34 = 3400$. You might think there are not many numbers you can rewrite as differences of squares; however, all odd numbers can — and so can all even numbers that divide 4. The real problem with these calculations is that you really need a "feeling" for numbers; that is, you need to develop a skill that mentally breaks up numbers into components that are easier to deal with. With practice, this will come as a second nature.

It might come as a surprise to you that speed square rooting numbers is easier than speed squaring! All you need to know is the table of squares of the numbers between 0 and 9:

The ten digits	The square of the digits
0	0
1	1
2	4
3	9
4	16
5	25
6	36
7	49
8	64
9	81

We'll begin with the square root of numbers between 0 and 10,000. This limits the trick to square roots that have one or two digits. We'll show you the procedure and follow along with a problem: Find the square root of 6,561. We'll refer to the answer to this problem as the "solution".

- $\sqrt{6561} = ?$

  - Locate — in the right-hand column of the table — the number(s) between 0 and 100 whose units digit is the same as the units digit of the given number.

    → 6561 → 1 or 81

  - The units digit of the solution is the square root of this number — the number in the same row in the left-hand column of the table. If there are two such numbers the units digit is one of these.

    → 6561 → 1 or 81 → The units digit is either 1 or 9

  - Erase the last two (rightmost) digits of the given number to get a new number we'll call the "left-over" number.

    → 65~~61~~ → 65 is the "left-over" number

  - Locate — in the right-hand column of the table — the first square number less than the "left-over" number.

    → The closest square number lower than 65 is 64.

  - The square root of this number — the number in the same row in the left-hand column — is the tens digit of the solution.

    → The tens digit of the solution is the square root of 64.
    $\sqrt{64} = 8$

  - We now know the tens digit of the solution, and if the units digit of the given number is not 0 or 5, we have two options for the whole solution. (If it *is* 0 or 5, then we already have the whole solution.)

    → $\sqrt{6561}$ is either 81 or 89

- Multiply the tens digit of the solution by the whole number that comes directly after it. If the "left-over" number is less than the product, the units digit of the solution is the lower of the two options. Otherwise, it is the higher.

  → $8 \times 9 = 72$. $65 < 72$ so the units digit is 1 and: $\sqrt{6561} = 81$

  Here are two more examples.

- $\sqrt{4225} = ?$

  - Locate — in the right-hand column of the table — the number(s) between 0 and 100 whose units digit is the same as the units digit of the given number.

    → $4225 \rightarrow 25$

  - The units digit of the solution is the square root of this number — the number in the same row in the left-hand column of the table.

    → $4225 \rightarrow 25 \rightarrow$ The units digit is 5

  - Erase the last two (rightmost) digits of the given number to get a new number we'll call the "left-over" number.

    → $42\overline{25} \rightarrow 42$ is the "left-over" number

  - Locate — in the right-hand column of the table — the first square number less than the "left-over" number.

    → The closest square number lower than 42 is 36.

  - The square root of this number — the number in the same row in the left-hand column — is the tens digit of the solution.

    → The tens digit of the solution is the square root of 36. $\sqrt{36} = 6$

    → The solution is: $\sqrt{4225} = 65$

- $\sqrt{3249} = ?$

  - Locate — in the right-hand column of the table — the numbers whose units digit is the same as the units digit of the given number.

    $\longrightarrow$ 3249 → 9 or 49

  - The units digits of the solution is one of the square roots of these numbers — the numbers in the same row in the left-hand column of the table.

    $\longrightarrow$ 3249 → 9 or 49 → The units digit is either 3 or 7

  - Erase the last two (rightmost) digits of the given number to get a new number we'll call the "left-over" number.

    $\longrightarrow$ 32~~49~~ → 32 is the "left-over" number

  - Locate — in the right-hand column of the table — the first square number less than the "left-over" number.

    $\longrightarrow$ The closest square number lower than 32 is 25.

  - The square root of this number — the number in the same row in the left-hand column — is the tens digit of the solution.

    $\longrightarrow$ The tens digit of the solution is the square root of 25. $\sqrt{25} = 5$

    $\longrightarrow$ We have two options for the whole solution. $\sqrt{3249}$ is either 53 or 57

  - Multiply the tens digit of the solution by the number that comes directly after it. If the "left-over" number is less than the product, the units digit of the solution is the lower of the two options. Otherwise, it is the higher.

    $\longrightarrow$ $5 \times 6 = 30$. $32 > 30$, so the units digit is 7 and: $\sqrt{3249} = 57$

The speed square rooting is based on two facts. The first is that when you square a whole number, the units digit of the result is equal to the square of the units digit of the number being squared. This can easily be seen by writing an example where I square a number using long multiplication. Look at the long multiplication form of 39^2, the units digit of 9×9, 1, doesn't add to anything and so appears as the units digit of the result.

$$\begin{array}{r} 39 \\ \times\ 39 \\ \hline 351 \\ 117 \\ \hline 1521 \end{array}$$

The second fact is that the square of a given 2-digit number will always be larger than (or equal to) the nearest multiple of 10 below it and lower than (or equal to) the nearest multiple of 10 above it. For example, $39^2 = 1521$ is larger than $30^2 = 900$ but smaller than $40^2 = 1600$. The tens digit (3) is, therefore, between $3^2 = 9$ and $4^2 = 16$. We said at the start that this trick works for numbers up to 10,000. Now that you know how it works, you can generalise to larger numbers, the only difference being that you now will have a hundreds digit to worry about — and perhaps even more digits. This only makes the second part of the algorithm more difficult — but we can make an intelligent guess for this part. Here's how to find the square root of 331,776.

- $\sqrt{331776} = ?$

  - Locate — in the right-hand column of the table — the number(s) between 0 and 100 whose units digit is the same as the units digit of the given number.

    $\rightarrow$ 331776 $\rightarrow$ 16 or 36

  - The units digit of the solution is the square root of this number — the number in the same row in the left-hand column of the table. If there are two such numbers the units digit is one of these.

→ 331776 → 16 or 36 → The units digit is either 4 or 6

- Erase the last two (rightmost) digits of the given number to get a new number we'll call the "left-over" number.

  → 331776 → 3317 is the "left-over" number

- Find the first square number less than the "left-over" number — this won't be in the table anymore, so we'll have to estimate it.

  → Since $50^2 = 2500$ and $60^2 = 3600$, we'll try 58. $58^2 = 3364$, which is just slightly larger than 3,317, so the correct answer is $57^2 = 3249$, and the closest square number lower than 3,317 is 3,249.

- The square root of this number makes up the leftmost digits of the solution.

  → The leftmost digits of the solution are: $\sqrt{3249} = 57$

  → We have two options for the whole solution. $\sqrt{331776}$ is either 574 or 576

- Multiply the leftmost digits of the solution (all digits except the units digit) by the number that comes directly after it. If the "left-over" number is less than the product, the units digit of the solution is the lower of the two options. Otherwise, it is the higher.

  → $57 \times 58 = 3306$. $3317 > 3306$, so the units digit is 6, and: $\sqrt{331776} = 576$.

Admittedly, as numbers get bigger, this algorithm becomes quite a mental task. The most problematic step is estimating the square root of a large number that isn't a whole square. There are mental methods that can do that, but that's beyond the scope of this book.

The great thing about this speed squaring is that it can be easily generalised for higher powers. In fact, it even gets easier! Take, for example, speed cube rooting. The cube of a number — raising a number to the power of 3 — means multiplying that number by itself three times; for example, the cube of 2 is: $2^3 = 2\times2\times2 = 8$, and the cube of 12 is $12^3 = 1728$. Conversely, the cube root is the reverse process: it involves determining which number, when multiplied by itself three times, results in the given number; so the cube root of 8, denoted $\sqrt[3]{8}$, is 2, and $\sqrt[3]{1728} = 12$. For speed cube-rooting, you need to remember the cube of the numbers between 0 and 9:

The ten digits	The cube of the digits
0	0
1	1
2	8
3	27
4	64
5	125
6	216
7	343
8	512
9	729

Notice that in this table, every units digit is unique, which makes things easier. Here's how we speed cube root, using 185,193 as an example. The procedure is nearly exactly the same as speed squaring, making only a few small adjustments.

- $\sqrt[3]{185193} = ?$

  - Locate — in the right-hand column of the table — the number between 0 and 1,000 whose units digit is the same as the units digit of the given number.

    → 185193 → 3

- The units digit of the solution is the cube root of this number — the number in the same row in the left-hand column of the table.

  $\longrightarrow$ 185193 $\rightarrow$ 3 $\rightarrow$ The units digit is 7

- Erase the last *three* (rightmost) digits of the given number to get a new number we'll call the "left-over" number.

  $\longrightarrow$ 185~~193~~ $\rightarrow$ 185 is the "left-over" number

- Locate — in the right-hand column of the table — the first cube number less than the "left-over" number.

  $\longrightarrow$ The closest cube number lower than 185 is 125.

- The cube root of this number — the number in the same row in the left-hand column — is the tens digit of the solution.

  $\longrightarrow$ The tens digit of the solution is 5, the cube root of 125.

  $\longrightarrow$ The solution is: $\sqrt[3]{185193} = 57$

Finding the 4th root of a number doesn't require a new table. All you have to do is to do the square root routine twice. For example:

$$\sqrt[4]{331776} = \sqrt{\sqrt{331776}} = \sqrt{576} = 24$$

Finding the 5th root of a number is considered the easiest, because the units digit of the solution is the same as the units digit of the problem! You can use the following table to do mental calculations of the 5th root. Mind you, remembering the ten numbers in the table is becoming harder and harder as the power grows.

The ten digits	The 5th power of the digits
0	0
1	1
2	32
3	243
4	1024
5	3125
6	7776
7	16807
8	32768
9	590499

Recap

We have barely covered anything, but I've used up so much space! I wish we could learn some speed division tricks, mental fraction arithmetics, or even some calendar magic — something I'm really into now. Still, I hope this chapter has whet your appetite, and you can find some great references in the bibliography below. We have learned much of the basics, though, and here are some of the main concepts we touched upon.

- *Speed math/Mental arithmetic* — the ability to perform mathematical calculations in your head.

- *Human calculators/Lightning calculators* — prodigies who can perform speed math.

- *Gauss' method* — a method that uses symmetry considerations for speed addition of an arithmetic series of numbers.

- *Vedic Mathematics* — a speed math book written by the Indian monk and mathematician Bharati Krishna Tirthathat in 1965 and a term now widely used to mean speed math; however, there is no known connection to the ancient Vedic scriptures.

- *Mathematical Recreations and Essays* — One of the first best-selling recreational math books of the 20th century written by Walter Rouse-Ball and Donald Coxeter.

- *Taxicab numbers* — the smallest number that can be written as a sum of two cubes in a given number of ways.

- *The Trachtenberg speed system* — a systematic approach to basic math calculations developed by Jakow Trachtenberg, founder of the Mathematical Institute in Zurich, Switzerland.

- *Left-to-right multiplication* — a method to numbers beginning with the left (most-significant) digit.

- *Speed addition, subtraction, multiplication, squaring, cubing, square-rooting, cube-rooting, etc.* — arithmetic, mental math.

Challenge Yourself!

(1) Mentally find the sums of the following expressions and explain your method:
   (a) All the odd numbers between 1 and 100
   (b) All the even numbers between 1 and 100
   (c) All the numbers between 1,234 and 5,678

(2) Mentally calculate the following percentages and explain your method:
   (a) 17% of 25
   (b) 64% of 125
   (c) 243% of $11\frac{1}{9}$

(3) Mentally calculate the following products and explain your method:
   (a) 54315987×12
   (b) 46×54
   (c) 613×23

(4) Mentally calculate the following squares and explain your method:

(a) 155^2

(b) 96^2

(c) 261^2

(5) Mentally calculate the following roots and explain your method:

(a) $\sqrt{2304}$

(b) $\sqrt[3]{704969}$

(c) $\sqrt[4]{28561}$

(d) $\sqrt[5]{6590815232}$

(6) Prove that every even number that is divisible by 4, and every odd number, can be written as the difference between two squares.

(7) Use the expression $(a+b)\times(a-b)=a^2-b^2$ to find the prime factors of the following numbers:

(a) 299

(b) 1189

(c) 9991

(8) Here's a problem that requires "out of the box" thinking. Find x if:

$$\sqrt{x+\sqrt{x+\sqrt{x+\sqrt{x+\ldots}}}} = 11$$

Hint: Amazingly, this can be done in your head!

Solutions

(1) (a) Using Gauss' method, we realise that: $1+3+5+\ldots95+97+99=(1+99)+(3+97)+(5+95)+\ldots$ and that equals: $100\times25=2500$.

(b) We could use the same method as in (a), or we could just subtract the sum of the odd numbers, 2,500, from the sum of all the numbers, 5,050, to get $5050 - 2500 = 2550$.

(c) $5678 - 1234 = 4444$, so there are $4444 + 1 = 4445$ numbers between 1,234 and 5,674 if we include both 1,234 and 5,678. Since this is an odd number, using Gauss' method, we have 2,222 times the sum: $1234 + 5678 = 6912$, and we need to add the middle number, $6912 \div 2 = 3456$, to get $2222 \times 6912 + 3456 = 15361920$.

(2) (a) 17% of $25 = 25\%$ of $17 = 17 \div 4 = 4.25$.

(b) 64% of $125 = 125\%$ of $64 = 125 \times 64 \div 100 = 64 \times 10 \div 8 = 80$.

(c) 243% of $11\frac{1}{9} = 243\% \times 100 \div 9 = 243 \div 100 \times 100 \div 9 = 243 \div 9 = 27$.

(3) (a) Use Trachtenberg's method. Add a leading zero to the number: 054315987. Write underneath each digit the sum: twice the digit plus its neighbour to the right to get: $54315987 \times 12 = 651791844$.

(b) Use the expression $(a + b) \times (a - b) = a^2 - b^2$ to get: $46 \times 54 = (50 + 4) \times (50 - 4) = 50^2 - 4^2 = 2500 - 16 = 2484$.

(c) Use left-to-right multiplication to get: $613 \times 23 = 600 \times 23 + 10 \times 23 + 3 \times 23 = 13800 + 230 + 69 = 14099$.

(4) (a) Use the "squaring numbers ending in 5" trick to get: $155^2 = 100 \times 15 \times 16 + 25 = 100 \times 15 \times 15 + 100 \times 15 + 25 = 22500 + 1500 + 25 = 24025$.

(b) Use the "squaring numbers close to 100" trick. Note the positive difference between the number and 100. $100 - 96 = 4$. Subtract this difference from 96. $96 - 4 = 92$. Multiply the result by 100. $92 \times 100 = 9200$, and add the square of the difference you found earlier. $9200 + 4^2 = 9200 + 16 = 9216$.

(c) We'll use a modified version of the "squaring numbers close to the nearest 100" trick. Note the positive difference between the number and 250. $261 - 250 = 11$. Add this difference to 261. $261 + 11 = 272$. Multiply the result by 250. $272 \times 250 = 272 \times 1000 \div 4 = 68000$, and add the square of the difference you found earlier. $68000 + 11^2 = 68000 + 121 = 68121$.

(5) (a) Using the square root routine: $\sqrt{2304} = 48$.

(b) Using the cube root routine: $\sqrt[3]{704969} = 89$.

(c) Using the square root routine, twice: $\sqrt[4]{28561} = 13$.

(d) $\sqrt[5]{6590815232} = 92$.

(6) There are many ways to prove that every even number that is divisible by 4 and every odd number can be written as the difference between two squares. I like this pictorial way, where we use a square grid to prove our point. Whenever we refer to a single 1×1 square, we'll just use the word "square"; otherwise, we'll specify its size or use the general label $n \times n$ square. First, we'll start with the odd numbers 1 and 3 and work our way up through the odd numbers. 1 is odd and $1 = 1^2 - 0^2$, so that works. For 3, we'll draw a 2×2 square, as shown in Fig. 3.8. We can colour the 1×1 square within it green. There are 3 remaining uncoloured squares: one from the row, one from the column, and one from the top corner. Hence, $3 = 2^2 - 1^2$. Generally, we can always colour an $n \times n$ square that is smaller by one row and one column from a given $(n + 1) \times (n + 1)$ square, and this will always leave an odd number of uncoloured squares because we are adding 1 — the corner square — to an even number — twice the length of the smaller $n \times n$ square. Doing this in sequence gives all the odd numbers in order, as shown in the top row in Fig. 3.8 for the odd numbers 3, 5 and 7.

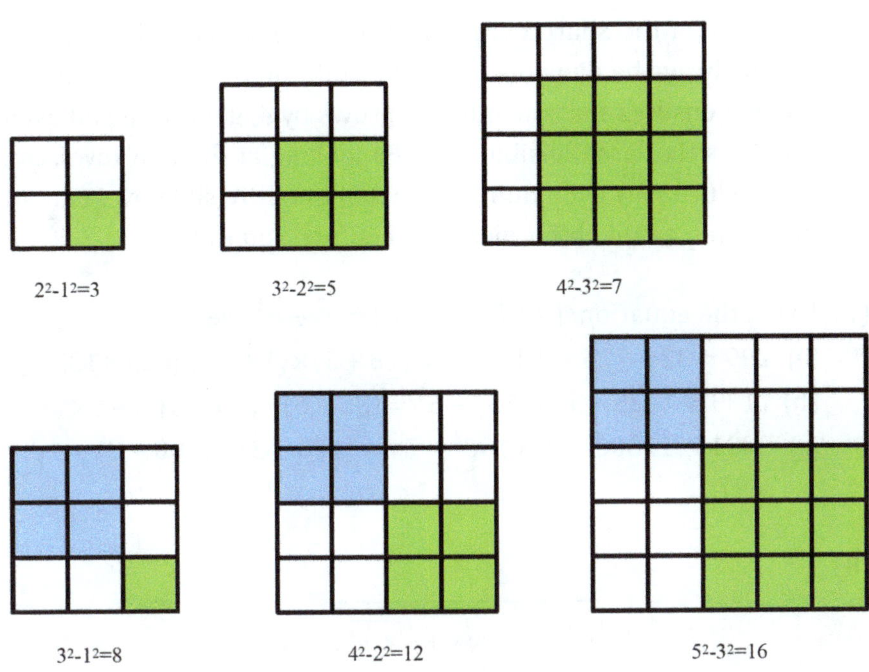

Fig. 3.8 Pictorial proof that every odd number and even numbers that are divisible by 4 can be written as the difference between two squares

We can use a similar procedure to generate even numbers that are divisible by 4 without remainder. $4 = 2^2 - 0^2$. Next is $8 = 3^2 - 1^2$. This can be depicted as a 3×3 square within which one square is green. We also colour the upper 2×2 square blue. Next is $12 = 4^2 - 2^2$. This can be depicted as a 4×4 square within which a 2×2 square is green. Again we colour the upper 2×2 square blue. The three even numbers that divide 4: 8, 12 and 16 are depicted in the bottom row of Fig. 3.8. The series can be continued by adding one row and one column, colouring the the upper 2×2 square blue in each case and the bottom corner square — whatever's left — green. The even number will be the number of blue coloured squares — always 4 — plus twice a rectangle (or square) that has at least one side of length 2 (because of the shared edge with the blue square). This is why it will always be divisible by 4. The addition of one row and one column ensures that the number grows by 4, generating all even numbers that are divisible by 4. By looking at the two rows, can you deduce why even numbers that are not divisible by 4 cannot be written as the difference between two squares?

(7) Using the equation, $(a + b) \times (a - b) = a^2 - b^2$, we get:

(a) $299 = 324 - 25 = 18^2 - 5^2 = (18 + 5) \times (18 - 5) = 23 \times 13.$

(b) $1189 = 1225 - 36 = 35^2 - 6^2 = (35 + 6) \times (35 - 6) = 41 \times 29.$

(c) $9991 = 10000 - 9 = 100^2 - 3^2 = (100 + 3) \times (100 - 3) = 103 \times 97.$

(8)

$$\sqrt{x + \sqrt{x + \sqrt{x + \sqrt{x + \ldots}}}} = 11$$

Square both sides of the equation to get:

$$x + \sqrt{x + \sqrt{x + \sqrt{x + \sqrt{x + \ldots}}}} = 121$$

Since we know from the problem that:

$$\sqrt{x + \sqrt{x + \sqrt{x + \sqrt{x + \ldots}}}} = 11,$$

we can plug that back in to get: $x + 11 = 121$, so: $x = 121 - 11 = 110$.

I find this a really neat way to solve a complicated equation!

Bibliography and Further Reading

Ball, W. W. R. and Coxeter, H. S. M. (1987). *Mathematical Recreations and Essays*. Dover Publications. 360–387.

Benjamin, A. and Shermer, M. (2006). *Secrets of Mental Math*. Three Rivers Press.

Dehaene, S. (2000). Geniuses and prodigies. *Math Horizons* **7**, 3: 18–21. http://www.jstor.org/stable/25678251

Devi, S. (1977). *The Joy of Numbers*. Harper & Row.

O'Connor, J. J. and Robertson, E. F. (1999a). Memory, mental arithmetic and mathematics. *MacTutor Index*. School of Mathematics and Statistics, University of St. Andrews, Scotland. Retrieved September 28, 2023, from https://mathshistory.st-andrews.ac.uk/HistTopics/Mental_arithmetic/.

O'Connor, J. J. and Robertson, E. F. (1999b). Srinivasa Aiyangar Ramanujan. *MacTutor Index*. School of Mathematics and Statistics, University of St. Andrews, Scotland. Retrieved September 28, 2023, from https://mathshistory.st-andrews.ac.uk/Biographies/Ramanujan/.

Trachtenberg, J., Cutler, A., and McShane, R. (2011). *The Trachtenberg Speed System of Basic Mathematics*. Ishi Press International.

Wikipedia contributors. (2023, September 2). Srinivasa Ramanujan. *Wikipedia, The Free Encyclopedia*. Retrieved September 26, 2023, from https://en.wikipedia.org/w/index.php?title=Srinivasa_Ramanujan&oldid=1173467610.

Wikipedia contributors. (2023, September 25). Carl Friedrich Gauss. *Wikipedia, The Free Encyclopedia*. Retrieved September 28, 2023, from https://en.wikipedia.org/w/index.php?title=Carl_Friedrich_Gauss&oldid=117707524.

Chapter **4**

Mazes

Introduction

Mazes never cease to amaze us! In this chapter, we'll explore the history of mazes, beginning with labyrinths and ending with some of the more quirky and unusual types. For starters, we'll begin with the classic maze. You go in and then have to find your way out.

The Puzzle

Solve the maze in Fig. 4.1.

Fig. 4.1   A classic maze

Where to Start?

This is a classic maze, so you shouldn't find it too difficult — we'll run into those later. One useful tip for solving mazes is to try and solve them backwards, from the exit to the entrance.

Solving the Puzzle

Figure 4.2 shows the solution to the maze.

Fig. 4.2   Solution to the classic maze

The History of the Puzzle and its Inventor

Mazes have amazed humanity since antiquity. Why do we like them so much? Most mazes are set out in front of you from the start, with rules that are easy to understand, making it easy to jump in and start solving them. Mazes come for all ages and abilities, so they have a wide reach. And they are visually stunning. Psychologists have suggested many other reasons. Like all other puzzles, they enhance creative thinking skills, so they're excellent for your brain. They also help boost confidence. When solved by hand, they enhance motor and visual skills, and when solved as a group, especially outdoor or indoor physical mazes, they can also build social skills like teamwork and leadership.

We know a lot about the history of mazes from William H. Matthews' excellent book, *Mazes and Labyrinths: A General Account of Their History and Development.* The book, published in 1922 and now in the public domain, is a real treasure.

What's the difference between a maze and a labyrinth? William Matthews proves that historically, the two terms have been used interchangeably, though rigorously speaking, there is a definitive difference. Labyrinths are a class of mazes with a single path leading to the maze's centre and no other paths at all. This sounds ridiculous at first. Isn't this just a cul-de-sac? We'll see in a few minutes that mathematically, at least, labyrinths can be very interesting.

Even Matthews, in a chapter he devotes to etymology, at least partly acknowledges the distinction between mazes and labyrinths that we adopt in this chapter. He writes:

> *"We might perhaps point out that a slight shade of difference may be assumed to exist between 'labyrinth' and 'maze,' even when these words are used in their metaphorical sense. We may take 'labyrinth' to sig-nify a complex problem involving merely time and perseverance for its solution, 'maze,' on the other hand, being reserved for situations fraught, in addition, with the elements of uncertainty and ambiguity, calling for the exercise of the higher mental faculties — in short, we may regard the two words as having reference respectively to the unicursal and multicursal types of plan. A distinction of this kind adds point to a sentence like that which occurs, for instance, in*

> *Mr. Lytton Strachey's 'Queen Victoria,' where he tells us (p. 178) that the Prince Consort 'attempted to thread his way through the complicated labyrinth of European diplomacy, and was eventually lost in the maze."*

The first record of a maze to date is the Labyrinth of Egypt. It was constructed by Amenemhat III about 4,000 years ago near a pyramid at a place called Hawara, south of Cairo. Greek and Roman historians and geographers rave about the place and applaud its ingenious complexity in their classical works. Herodotus, a 5th century B.C. author, even goes as far as writing that it surpasses even the pyramids! Though destroyed in antiquity, modern archaeologists and historians have tried to reconstruct a map of it from the descriptions of Herodotus and other classical authors. One such reconstruction, by the 19th-century Italian archaeologist Luigi Canina, is shown in Fig. 4.3. To date, excavations have uncovered only a few remnants of the foundations of the maze.

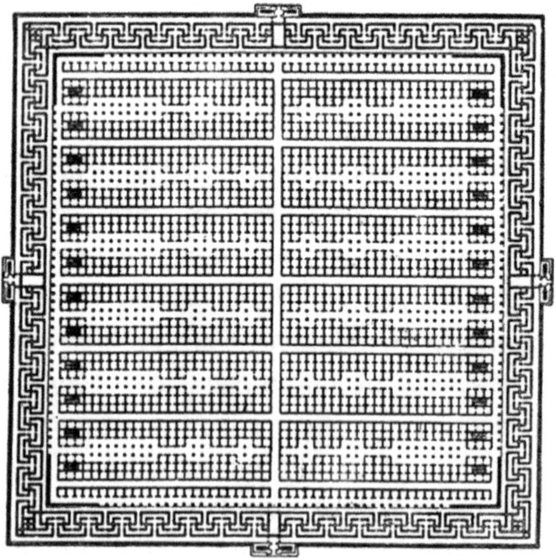

Fig. 4.3   Possible restoration of the Labyrinth of Egypt by Italian archaeologist Luigi Canina

Possibly the most well-known maze is the Cretan Labyrinth, home to the infamous Minotaur. There is considerable controversy about where this labyrinth was, when it was built, and what it looked like. We do know that it was already on the island of Crete, Greece, in the Bronze Age because of its distinctive pattern (Fig. 4.4), sometimes including the half-man, half-bull image of the Minotaur that appears on many antique artefacts from that era.

Fig. 4.4   The classic Cretan Labyrinth design

The labyrinth's fame is due to the Greek legend of Theseus and the Minotaur. To cut a long story short, King Minos punished the Athenians for killing his eldest son, Androgeus, by forcing them to send every few years — different sources have different views on how often this happened — the seven most courageous boys and the seven most beautiful girls to the Island of Crete as an offering to the Minotaur, a half-man, half-bull monster that lived in the middle of the Cretan Labyrinth. This always ended up in them becoming his dinner. That is, until Theseus, a legendary warrior, cunningly replaced one of the boys, hid a sword beneath his tunic, and when his turn came, chopped the Minotaur's head off. He escaped the labyrinth with the aid of a ball of thread, one end of which he tied to the entrance and the other he kept trailing along with him to mark the way in — and out (why didn't Hansel and Gretel think of that?). The thread was given to him by Ariadne, one of the seven girls who fell in love with him. Unfortunately for her, he dumped her on another island on the way back to Athens. Ariadne wasn't the only thing he forgot. He also forgot his promise to his father, Aegeus, who just happened to be the king of Athens, to erect a white sail on the way back if his mission succeeded. The sail was black; Aegeus got depressed — thinking the monster killed his son — and promptly committed suicide by jumping off the cliff into the sea, suitably known today as the Aegean Sea. Edward Burne-Jones, a British painter and designer, captured the moment Theseus and the Minotaur first saw each other in the 1861 portrait *Theseus in the Minotaur's Labyrinth* (Fig. 4.5). The portrait is marvellous. I love the way Theseus is creeping up on the monster. You can see the bones of the previously devoured victims, the ball of thread in Theseus's hand, the anticipation and perhaps slight apprehension on the Minotaur's face, and the labels: "Theseus", "Minotaurus" and "Labyrinth", just to make sure we all know who's who.

The Cretan Labyrinth design became widespread and is a good example of a real labyrinth, a maze with a solitary path to the centre. Even though the solution is absolutely trivial — just walk straight, and you'll get to the centre — the many twists and turns of the path might confuse and disorientate someone, and if the walls are high, you'd certainly get claustrophobic.

The Cretan Labyrinth is an example of what mathematicians call a *simple, alternating transit maze*, also known by its acronym, an *SAT maze*. Transit means there is one path to the maze's centre. The words "simple" and "alternating" relate to the maze's specific

Fig. 4.5   Edward Burne-Jones painting of *Theseus in the Minotaur's Labyrinth*

structure. Looking at the labyrinth from above, you can see its concentric layered arrangement. Each "circle" or layer is called a level, and it is labelled by a number, 0, 1, 2... *n*, according to its position relative to the outside of the maze. So, the outside of the maze is labelled 0, the first circle (level) 1, the next 2, and so on, until the maze's centre is labelled *n*. Figure 4.6 shows the labelling of the levels superimposed on the Cretan Labyrinth design. Start outside the labyrinth and go in. You go directly from level 0 to level 3 and turn right, going anti-clockwise along the full length of the circular path until it curves right into level 2, along which you are now travelling clockwise. At the end of level 2, you turn left, this time into level 1. You are now travelling anti-clockwise. The SAT

Fig. 4.6   The levels in the Cretan Labyrinth

maze is an *alternating maze* because you alternate between travelling clockwise and anti-clockwise, left and right, everytime you change levels. It is a *simple maze* because you visit each level exactly once, never to return. In a simple maze, you could not go, for example, from level 6 to level 1 and then to level 3, level 4 and back to level 1 again. Let's write down the levels in the Cretan Labyrinth as we travel from outside the maze to its centre. 0, 3, 2, 1, 4, 7, 6, 5, 8. This sequence is known as the level sequence of the SAT maze.

Figure 4.7 shows another SAT, known as the Jericho Labyrinth. It illustrates the 14th-century Farhi Bible, an illuminated manuscript by the Majorcan Jewish cartographer Abraham Cresques. Cresques imagined that the seven walls of Jericho, brought down by Joshua, who had circled them seven times while shouting and blowing trumpets, *had* to be built as a concentric labyrinth. The level sequence of the Jericho Labyrinth is 0, 3, 4, 5, 2, 1, 6, 7.

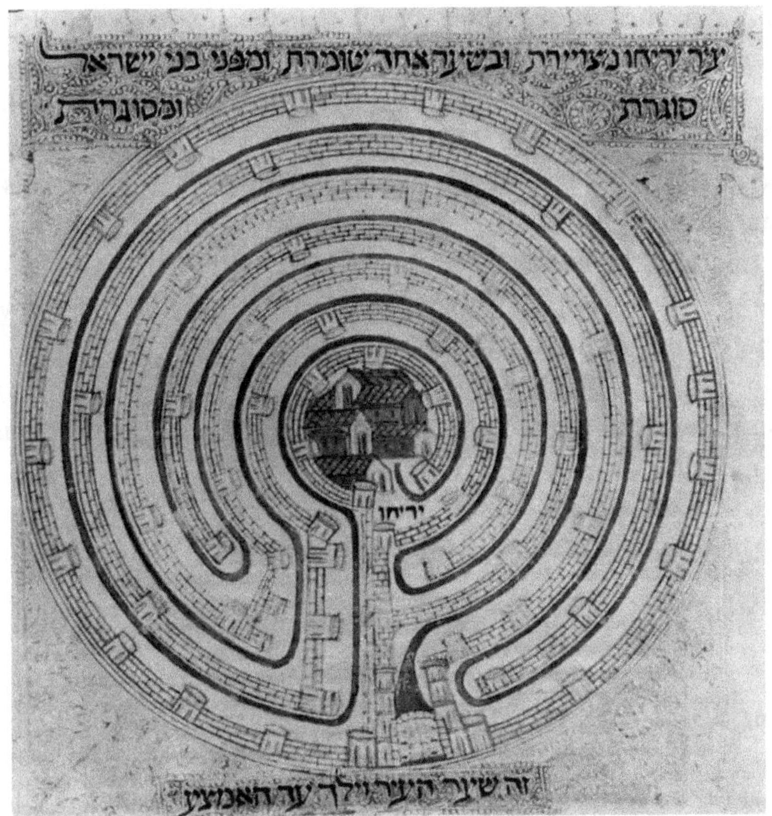

Fig. 4.7   The Jericho Labyrinth

Tony Phillips, a mathematician at Stony Brook University in New York, has a great website that explains a lot about SAT mazes and some of their interesting mathematical properties, including their three main properties. First, they all begin with 0 and end with n because they indicate the layers a person encounters on his journey from outside the maze (0) to its centre (n). Second, they also always alternate odd and even numbers. Here's a nice *proof by contradiction.*

Proving by contradiction means that you make the opposite assumption to the given statement and show that you inevitably reach a logical fallacy. Suppose a level sequence contains two adjacent even numbers or two adjacent odd numbers (this is called having the same *parity*). This means that when you go through the maze, at some point, you go directly from an even level to another even level or directly from an odd level to an odd level. Regardless of whether they are both even or both odd, by definition, there has to be an odd number of levels between them because the levels are numbered *sequentially* from the outside, 0, inward to *n*. Stating that two adjacent levels in a SAT sequence can both have the same parity, therefore, necessitates an odd number of levels nested between them. It is this fact that we are going to contradict.

In any SAT maze, levels that are nested in a region between two adjacent levels in the sequence have to enter and exit that region on the opposite side of where the vertical transition between the two levels is made. This is illustrated in the Cretan Labyrinth in Fig. 4.8.

Fig. 4.8   Levels nested between two adjacent sequence levels must enter and exit from the side opposite to the vertical transition path between the two adjacent levels

Levels 1 and 4 are nested. The short vertical path joining them is on the left, and the nested levels 2 and 3 enter and exit from the right. But this means there *has* to be an even number of levels between two adjacent sequence levels, which contradicts the fact that if the two adjacent levels have the same parity, there must be an odd number of levels between them.

The third main property of SAT-maze sequences applies to any consecutive pair of numbers that begin with the same parity. If they overlap, like the pairs $(1, 4)$ and $(3, 2)$, then one is nested within the other. To summarise, the three properties of SAT sequences are:

- they begin with 0 and end with n,

- they consist of alternating odd and even numbers, and,

- pairs of numbers that begin with the same parity are nested.

These completely and uniquely determine the structure of a SAT maze. This means that each SAT sequence describes exactly one SAT maze — or two — if we count mirror images as a separate maze. Given any SAT sequence, you will know all there is to know to draw its maze.

You might be curious about how many SAT sequences there are for any given n. Funnily enough, mathematicians have not yet found a formula that answers this, and they have to rely on computers to go through all possibilities. To date, we know that there's one maze for $n = 1$, 2 and 3, two for $n = 4$, three for $n = 5$, eight for $n = 6$, 14 for $n = 7$, 42 for $n = 8$, and then the numbers become really big. For a 15-level sequence, there are 30,694 mazes, and for $n = 55$, the last calculated number to date (as far as I know), there are a whopping 10,726,008,363,361,842,734,385,644 SAT mazes! How do I know these numbers? I looked them up in the best reference there is for number sequences, the Online Encyclopaedia of Integer Sequences (OEIS). The OEIS is a massive endeavour created by British-American mathematician Neil Sloane. Founded in 1964 and with over 375,342 sequences, it is full of treasures. All you need to do is go online — the link is in the bibliography section — type in the first few numbers of a sequence — and bang! You get all the information known about the sequence. Alternatively, you might want to invent your own sequence and submit it to the database. The SAT maze count is sequence A005316 in the encyclopaedia, but if you look there, you'll find that it says nothing about mazes! The sequence is called: "Meandric numbers: number of ways a river can cross a road n times". That's because the problem of finding the number of SAT mazes turns out to be exactly the same problem as counting the number of ways a river can cross a road. Two problems that have the same underlying mathematical structure are said to be *isomorphic*, which means that they are equivalent. Formally, a *meander* is a self-avoiding curve that crosses a straight line a number of times, hence the analogy of the river (self-avoiding curve) crossing the road (the straight line). Figure 4.9 shows an $n = 5$ open meander — "open" means that the beginning and end of the self-avoiding curve don't join. The meander sequence is created by assigning a number beginning with 1 to each part of the river, increasing it by one each time it crosses the road, and then reading off the numbers from left to right along the road. The sequence for

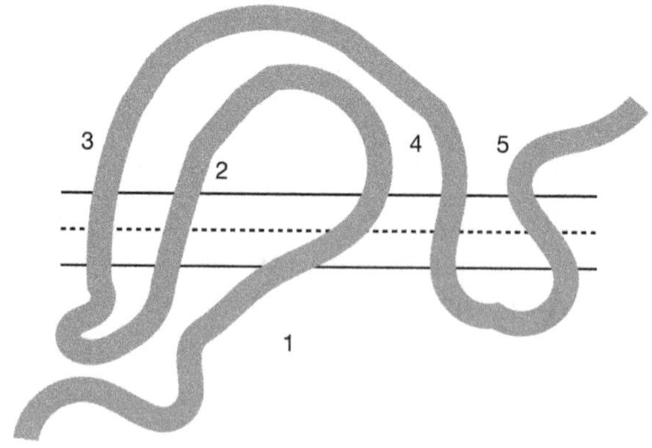

Fig. 4.9   An $n = 6$ meander

the meander in Fig. 4.9 is 3, 2, 1, 4, 5. Add a zero at the beginning and a 6 at the end — for start and end points — and you get the equivalent SAT maze sequence for $n = 6$: (0, 3, 2, 1, 4, 5, 6), which is the maze in Fig. 4.10.

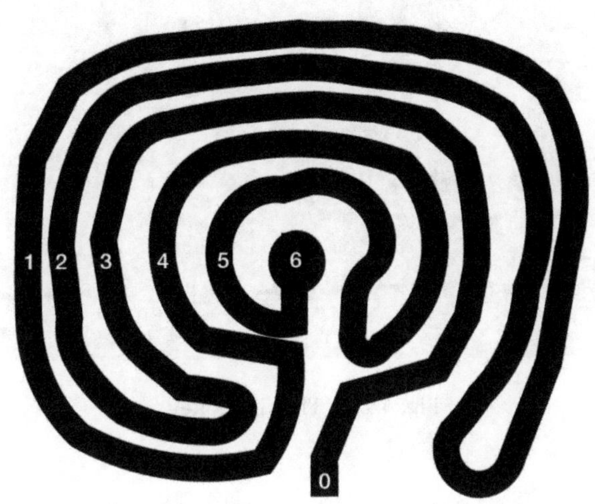

Fig. 4.10 The (0, 3, 2, 1, 4, 5, 6) SAT maze

The word meander comes from the Meander River, mentioned in Homer's *Iliad* as one of the places the Carians — a Troyan ally — came from. The river, now known as Büyük Menderes River, flows from Dinar in Southwest Turkey down to the Aegean Sea. Curving and winding along the way, it fittingly lends its name to the term "meandering". Today, "meander" is also used to describe a decorative border made from a single line that forms a repeating pattern, like the Greek key in Fig. 4.11. Note the similarity between these meanders, the river-meandering-across-a-road problem, and labyrinths.

If you want to learn more about meanders, you can read an article about meanders, knots, labyrinths, and mazes that was written by New Jersey mathematician Jay Kappraff and Serbian mathematicians Ljiljana Radović and Slavik Jablan and presented

Fig. 4.11   The Greek key

together with some of their mathematical labyrinths at the Bridges conference, a conference dedicated to the interconnections between math and art. The paper was also published in the *Journal of Knot Theory and Its Ramifications.*

Labyrinths are abundant throughout history — as walk-through mazes, decorations on coins, pottery, and other artefacts, in manuscripts, and also on Roman mosaic pavements (Fig. 4.12) and in churches (Fig. 4.13).

What about mazes — and I'm using the term to describe what we call mazes nowadays, with branchings and dead ends. It is difficult to pinpoint when these came into being, but until the Renaissance, they were much less popular and abundant than labyrinths. However, things changed in the Renaissance when hedge mazes became popular. One of the most famous mazes, and the oldest surviving one in Britain, is the Hampton Court maze, just outside London. King William III commissioned two gardeners to do the job, Henry Wise and George London, and they finished it in

1695. Figure 4.14 shows the layout of the maze. Since then, mazes have become more popular and have taken the lead in the labyrinth vs. maze competition.

Fig. 4.12 The Theseus mosaic

Fig. 4.13 A labyrinth in Grace Cathedral, San Francisco

Fig. 4.14 Layout of Hampton Court maze

The history of maze solving is far more recent, although the "hand-on-wall routine" must have been known since mazes were invented. The method goes as follows. Put your hand on the wall of the maze and never let go. So if the wall turns to the right, you go with it, and so on. Foolproof. You're guaranteed to get to the exit — unless the maze is not *simply connected*. A simply connected maze is one where all the walls are connected. Obviously, if it is not simply connected, for example, there are some disjointed walls — perhaps a "loop" within the maze — then you might find yourself going around in circles. The other problem with the hands-on-wall rule is that, although it guarantees success, it doesn't guarantee the quickest route. For mazes that have many paths, this might not be the best routine to follow. Still, it has one big advantage — and that is that you don't need to know the structure of the maze. So if you're on an outing and stumble across a random hedge maze, like Hampton Court's, with no prior knowledge or map of the maze, it can be a very useful way not to get lost.

Maze-solving techniques can be divided into those that require prior knowledge of the maze structure and those that don't. The "hands-on-wall routine" doesn't require prior knowledge. The "random mouse routine" doesn't either, though you might argue that this isn't a technique at all. All you need to do is, at every junction in the maze, make a random choice until you solve the maze. Recently, a paper written by Kalle Timperi, Alexander LaValle, and Steven LaValle, researchers from the University of Oulu in Finland, contains a (yet-to-be-established) proof that a robot moving pseudo-randomly in any maze will always eventually solve it. However, the "pseudo-random" moves of the robot have to be generated in a very special way — from the base 4 Champernowne constant, if you have to know (we won't go down that alley here).

Although the random mouse rule sounds as stupid as you can get, it can be modified slightly to make one of the most popular computer algorithms used today. All you need to do is, when you get to a dead end, go back to the previous junction, mark your path as a "forbidden entry", and make a new random choice. This method is known as "backtracking", a general term that describes techniques where you traverse a network of potential paths to solutions, make a choice at each decision junction along the way, and upon reaching an invalid solution, discard it and return to the previous decision junction to make a new, different choice. Backtracking methods were invented by computer scientists in the 1950s — it is unclear who exactly first coined the term — and used to solve puzzles like the eight queens puzzle and, later on, chess, Sudoku, and many other problems. However, these methods must have been used before computers by those who solved puzzles meticulously by re-tracking their steps when reaching dead ends.

In fact, we know of one backtracking method used for solving mazes and similar structures in the 19th century that is still very popular today. Invented in the 19th century by French telegraph engineer Charles Pierre Trémaux, this method is known today as "Trémaux's algorithm" or the "depth-first search" algorithm — the basic idea is to travel as far into the maze as possible and then backtrack. Here's how you implement the algorithm in a maze. Imagine the maze as an indoor web of corridors where there is a door at the entrance and exit of each corridor. Whenever you go through a door, make a mark on the floor. Whenever you reach a junction, and only if the door you came through is marked, choose any other door. If all other doors are marked, go back through the door you just came through unless it is marked twice. Otherwise, choose the door with the least marks. Never go through a door with two marks. This algorithm will guarantee that you solve the maze. The maze-solving page in Wikipedia has a great animation of this maze-solving technique.

The "dead-end algorithm" is a way to solve a maze if you know its structure. Once the maze is in front of you, find all the dead ends and "fill them in" up to the first junction. This will leave you with the solution.

The structure of mazes is really important and interesting, but to get a hold of them, we first need some kind of abstract representation. Just as mathematicians represent labyrinths with SAT sequences, they represent mazes using graphs and graph theory. We'll sidestep a bit to explain graph theory in slightly more detail. Mathematical graphs *are not* the graphs you are used to — like x-y graphs, pie charts, or the like. A (mathematical) graph is simply a collection of points, called vertices, connected by lines, called edges, which represent relationships or connections between the points. There are many kinds of graphs. A simple graph is one where there are no loops — vertices connected to themselves — and no more than one edge between any two vertices. A complete graph is one where all the vertices are connected to each other. A disconnected graph is one with detached regions, such as graphs that have solitary vertices. A directed graph is one where the edges have a direction indicating a one-way relationship between two points. A cyclic graph is one where there is at least one cycle — a path that starts and ends at the same vertex; a cycle graph — not to be confused with the cyclic graph — is one where the whole graph is a single cycle, and an acyclic graph has no cycles at all. A tree is an acyclic graph where all the vertices are connected, but there is a unique path between any two. A planar graph is one whose edges intersect only at their endpoints and do not cross each other. In other words, it's a graph that you can draw on the plane. You can see examples of all these graphs in Fig. 4.15. They are all planar graphs (we won't discuss non-planar graphs in this chapter).

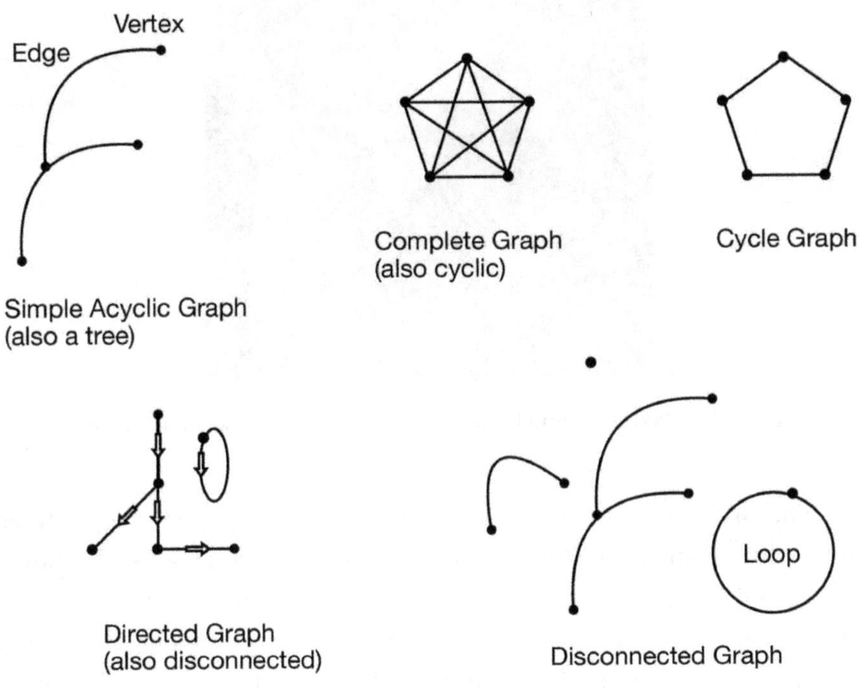

Fig. 4.15　Some graphs

The pioneer of graph theory was the 18th-century Swiss mathematician Leonhard Euler (Fig. 4.16). You might remember him from *Lewis Carroll's Cats and Rats ... and Other Puzzles with Interesting Tails*. There, we discussed him in the totally different context of Graeco-Latin squares. But his mathematical interests were far broader than that. He started off the mega-field of graph theory by solving a very practical problem known as the seven bridges of Königsberg.

Fig. 4.16   Jakob Emanuel Handmann's rendering of Leonhard Euler

The problem concerns seven bridges that span the Pregel River in the Prussian (now Russian) city of Königsberg (Fig. 4.17). Euler was asked to find a way to visit all parts of the city, crossing over all bridges but without crossing a bridge twice. Euler went about this using an ingenious method. He first replaced the over-detailed map with what we now call a (mathematical) graph. He represented the bridges as edges and the four unconnected patches of land as vertices. Figure 4.18 shows the graph. Once he did this, he found a rule that holds for all connected graphs.

The rule is that a path on a connected graph that visits every edge exactly once exists only on graphs where either each and every vertex has an even number of edges connected to it, or they have exactly two vertices that have an odd number of edges connected to them. The number of edges connected to a vertex is called the vertex's *degree.* The path that traverses all the edges is called an *Eulerian cycle* if the path begins and ends at the same vertex, and that happens if and only if the degree of all the vertices is even; it is called an *Eulerian trail* if the graph has exactly two odd vertices, in which case the path begins at one of them and ends at the other.

Fig. 4.17 Historical map of Königsberg with the bridges highlighted

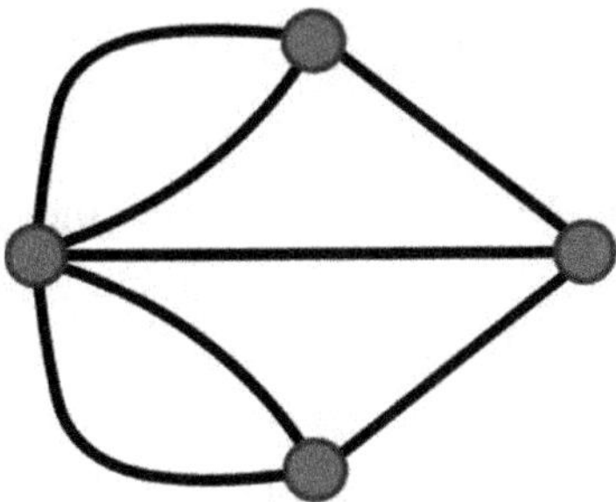

Fig. 4.18 Königsberg bridge problem as a graph. Edges represent bridges; vertices are the unconnected patches of land

A more formal way of putting Euler's theorem is: An Eulerian cycle exists on connected graphs with all-even vertex degrees, and an Eulerian trail exists on those with two odd vertex degrees. The vertex degrees of the Königsberg graph are all odd — they are 3, 3, 3 and 5 — hence, no Eulerian path (path meaning trail or cycle) exists. The Königsberg puzzle reminds me of similar puzzles that ask if you can draw a given image without taking your pencil off the paper. Can you draw the house on the left of Fig. 4.19? No — because four of its vertices have odd degrees. But you can draw the house on the right! Just reconfirming here that you *are* allowed to visit vertices more than once. The one-time visit only concerns the edges.

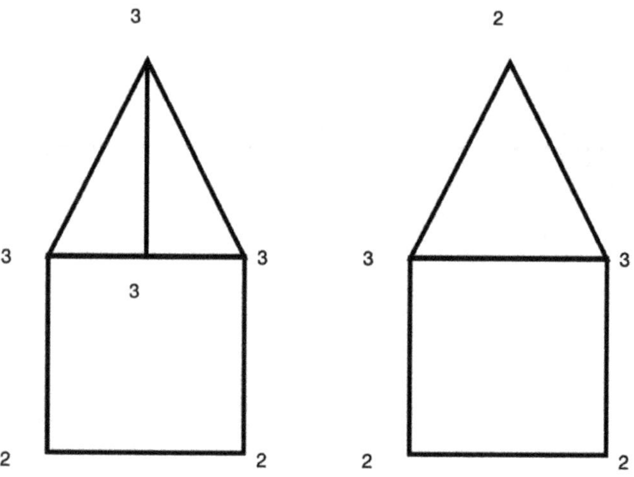

Fig. 4.19   An Eulerian path exists only on the house on the right

This immediately brings up the complementary problem — which graphs allow paths that visit each vertex exactly once (but you can travel along each edge any number of times)? Such a path is called a *Hamiltonian path*, and if it starts and ends at the same vertex, it is called a *Hamiltonian cycle*. Unlike Eulerian paths, there is no known method to decide *whether* a Hamiltonian path exists on a graph — and finding them can be tricky. It is also worth mentioning that *counting* the number of Eulerian or Hamiltonian paths on a graph is not easy, and as the graph gets larger, it becomes a very difficult, time-consuming computational problem, up to the point that it becomes prohibitive.

Hamiltonian paths are named after the 19th-century Irish mathematician, physicist, and Royal Astronomer William Rowan Hamilton (Fig. 4.20).

Fig. 4.20   William Rowan Hamilton

Hamilton was a child prodigy and a mental calculator — we mentioned him in Chapter 3 — and was even pitted against Zerah Colburn (he lost). He knew over a dozen languages, many of them non-European, like Sanskrit and Malay. In 1823, at the age of 17, he began his formal education — in mathematics and the Classics — at Trinity College, Dublin, coming first in every exam and in all topics. Throughout his years in Dublin — indeed, throughout his life —

alongside his career in math and physics, he retained his interest in languages and literature, regularly read and wrote poetry, and met and corresponded with many prominent poets of his day, including William Wordsworth and Samuel Coleridge.

In 1827, he was appointed Andrews Professor of Astronomy and Royal Astronomer of Ireland before he even finished his undergraduate degree! He took up residence at Dunsink Observatory just outside Dublin. Hamilton, an orthodox Christian, married Helen Marie Bayly, the daughter of a country preacher. On October 16, 1843, he was walking with his wife along Dublin's Royal Canal when he thought up the following equation: $i^2 + j^2 + k^2 = ijk = -1$ that described the multiplication rule for a new 4-dimensional complex number system he called *quaternions*. It was a problem that he had been working on for many years, and he was so excited by this that he took his penknife and actually carved the equation on the Brougham Bridge in Dublin as he was crossing. There's a plaque commemorating the occasion under the bridge. Hamilton was probably one of the first math graffiti artists! Quaternions are one of Hamilton's main contributions to math. But he made many more, not only in math but also in astronomy and physics. In fact, he laid down the mathematical foundations for much of modern physics, including optics, classical mechanics, quantum mechanics, electromagnetism, Fourier analysis, and linear algebra.

Ironically, many mathematicians studied Hamiltonian paths before they were named after him, and the interest in them dates back to 9th-century Indian and Islamic mathematicians. Among other things, they were interested in finding *knight tours* on chessboards, that is, finding a path that visits every square on a chessboard exactly once while moving like a knight: one square horizontally and two vertically, or two vertically and one horizontally in any direction (Fig. 4.21). Knight's tours are mathematically interesting, aesthetically pleasing, and great fun, and they have earned themselves a place of honour in recreational math. We've already seen

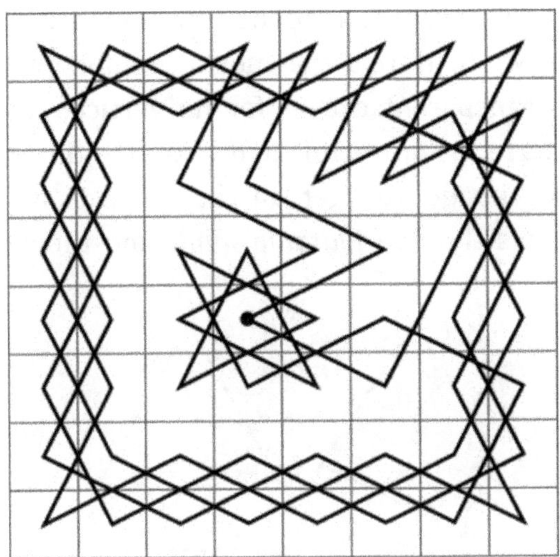

Fig. 4.21   A knight's tour on a chessboard

how they've been used in cryptography in Chapter 1. The knight's tour problem is equivalent to finding a Hamiltonian path on a 64-vertex graph, where each vertex represents one square on the chessboard, and each edge represents a knight's move, such that two vertices are connected with an edge if and only if there exists a knight's move between them. Just in case you're contemplating finding all the knight tours on a regular 8×8 chessboard — don't! There are 19,591,828,170,979,904 of them, of which 26,534,728,821,064 are closed (i.e. cyclic)!

So why did Hamilton get the honour of having these paths named after him? Most probably through a game he invented and rather unsuccessfully tried to commercialise under the name *The Icosian Game*. The aim of the game, which could be played by one or two players, was to find a Hamiltonian cycle on the face of a regular dodecahedron. It turned out that if the 12 vertices are labelled so that symmetrical solutions aren't equivalent, there are thirty different solutions. And it was pretty easy to solve, so it didn't become a popular puzzle. Hamilton's 3-dimensional dodecahedron, indeed all polyhedra, can be mapped into a planar (2-dimensional) graph in many ways. Here's the most common way. Imagine the polyhedron is transparent except for its vertices and edges. Now, pretend you have a powerful light source directly above the polyhedron. The image you get on the plane is its planar graph. Figure 4.22 shows the five *Platonic solids* and their Planar graphs.

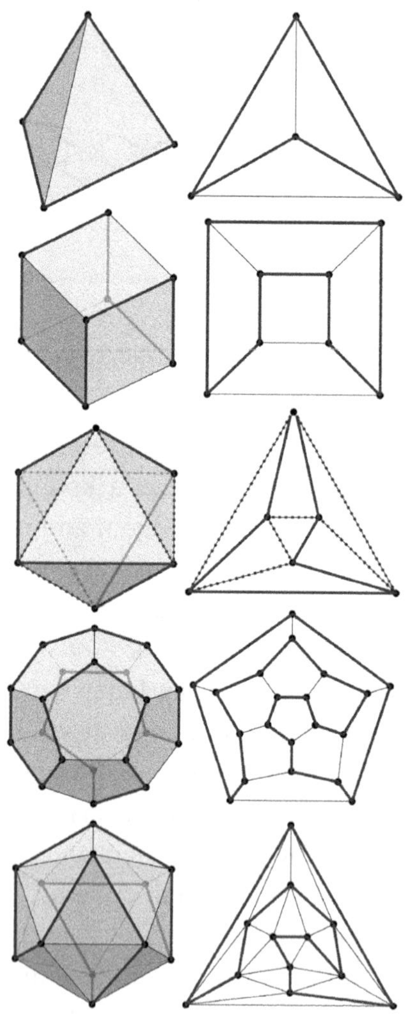

Fig. 4.22   Hamiltonian paths on the Platonic solids and their planar graphs

A Hamiltonian cycle is also shown for each of them. The five *Platonic solids* have been known since antiquity, but they are named after Plato, the Greek philosopher and mathematician, who mentioned them in one of his dialogues. The faces of the Platonic solids are regular polygons. The tetrahedron, dodecahedron and icosahedron's faces are regular triangles, while the cube and the octahedron are made from squares. In each Platonic solid, all the angles and all the edge lengths are equal, and the same number of faces meets at each vertex. They are the only 3-dimensional convex solids with these properties.

When mapping the Platonic solids to the graphs, the vertices are mapped to vertices and the edges to edges, so the numbers of vertices and edges are the same in the graphs as they are in the solids. What about the faces? The faces in the graph are the regions bounded by edges, but there is one less than in the corresponding solid. That is because the last face is projected as the area "outside" the graph (which is the same as saying that this is the face that is "flattened" when moving from three dimensions to two.

Table 4.1 shows the number of vertices, edges and faces for each of the Platonic solids. As you gaze at these numbers, you might discover an interesting phenomenon. The number of faces plus the number of vertices minus the number of edges is always 2, no matter which solid it is. This formula was first found by Sicilian mathematician Francesco Maurolico but is named after the mathematician who generalised it for all convex polyhedra, *Euler's formula*. Yes! The one and only, Leonhard Euler. It was generalised again by other mathematicians later on, but the name "Euler's formula" stuck, and so did the term *Euler's characteristic*, which is the number describing the overall shape of the polyhedra (the topological shape). For the Platonic solids, indeed, for all polyhedra with an overall spherical shape, and for all connected planar graphs, it equals 2 and is the sum of the vertices and the faces minus the edges, as we have seen.

Table 4.1   The number of vertices, edges and faces of the Platonic solids

Name	Vertices	Edges	Faces
Tetrahedron	4	6	4
Cube	8	12	6
Octahedron	6	12	8
Dodecahedron	20	30	12
Icosahedron	12	30	20

Graph theory is one of the most important, exciting and ever-developing fields in math, with many efficient mathematical tools for all kinds of analysis we might want to do. We've already seen how some things can be mapped onto graphs, so it won't be a big surprise to learn that mazes, too, can be mapped onto graphs and vice versa. The usual mapping is as follows: The vertices are all the places where decisions are made, i.e. path intersections and dead ends. The edges are the paths connecting them. Graph terminology is sometimes used to define different kinds of mazes. For example, a *simply connected maze* has no loops or circuits, and there is exactly one path from entry to exit. It is also called a perfect maze. A purely *multiply-connected maze*, also called a *braid maze*, is one that has circuits but no dead ends. A *unicursal maze* is a labyrinth, a maze without junctions. An *arrow maze* is a directed graph where, when you land on an arrow, you can move to any of the cells in either of the row, column or diagonal that it points to. Figure 4.23 shows a simple arrow maze with its corresponding graph. You have to get from the entry to the exit following the arrows. The entry and exit cells are those with the shaded background.

Today, you can find mazes everywhere. Prominent maze designers like Adrian Fisher construct all kinds of life-size mazes for public and private spaces, while others, like artist Elizabeth Carpenter, use special techniques to combine artwork and mazes that can be solved with pen and paper.

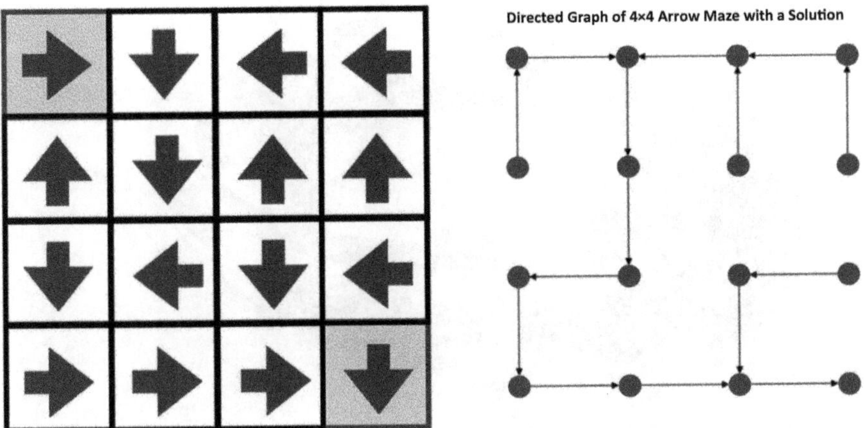

Fig. 4.23 A simple arrow maze and its corresponding directed graph

Generalisations of the Puzzle and Connections to Other Math Areas

The last few decades have seen an abundance of mazes with many features and scores of names. These generalisations involve changing some properties of the classic maze. For example, we've just seen that different mazes can be created according to their isomorphic graphs, like the simply- and multiply-connected mazes, the unicursal maze, and the arrow maze. But there are many other properties that we can change as well, like its rules or its topology. Here are some examples.

We can change the topology of a maze by taking it out of 2-dimensional space, either by making a 3-dimensional maze, adding height as an extra dimension, or by wrapping the 2-dimensional maze around a surface embedded in 3-dimensions, like a Möbius strip, a doughnut, a cube or a sphere. Figure 4.24 shows one such maze on a torus (a doughnut), and Fig. 4.25 shows the same maze but flattened out, so you can try and solve it if you wish (it's not too difficult). These "flattened-out" mazes are represented by drawing them like regular mazes, but you should imagine

Fig. 4.24   A maze on a torus

Fig. 4.25   A "flattened-out" torus maze

that the top edge connects to the bottom edge, and the left edge connects to the right edge, forming a continuous loop both in the up-down direction and in the left-right direction. Start at the "S" and end at the "F".

Most mazes are rectangular, where the paths intersect at right angles. They are usually designed on a square or rectangular grid, which makes things easier when using computer software to design mazes. The walls of the maze are cells that are coloured black, whereas the path is formed from the remaining uncoloured or white cells. *Rectangular mazes* are also called *gamma mazes. Triangular mazes,* also known as *delta mazes,* use equilateral triangles as the basic cells from which they are constructed, which lead to the paths meeting at angles that are multiples of 120°. Similarly, the basic cells of *hexagonal (sigma) mazes* are, well, hexagons. And, of course, any other basic cell shape can be used, including combinations, say, of octagons and squares (*upsilon*), mazes made from concentric circles (*theta*), or even a chaotic jumble of shapes (*crack maze*). You can see some of these mazes in Fig. 4.26.

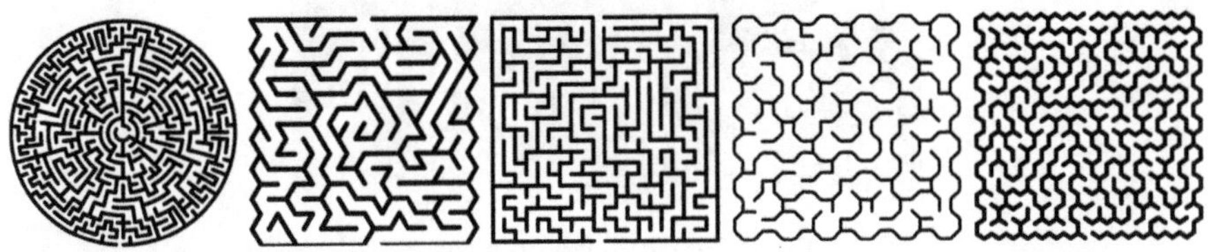

Fig. 4.26   From left to right: a theta maze, a delta maze, a gamma maze, a sigma maze, and an upsilon maze

Perhaps the most fun way to generalise the concept of a maze is to change the rules, like the arrow maze that we've already seen. Computer scientist and game inventor Robert Abbott (Fig. 4.27) was one of the most creative maze rule-changing "gurus". Celebrated by popular math author Martin Gardner in his mathematical games columns in *Scientific American,* he invented many intriguing mazes and coined the generic term for mazes with additional rules: *logic mazes.*

Any attempt to give a comprehensive review of even a small number of logic mazes is futile. There are just *so* many. And in any case, anyone can invent their own. So, I'll just show you three handpicked logic mazes so that you get the idea. The first is a very simple variation. The only difference from a classic maze is that you are only allowed to make left turns. Figure 4.28 shows such a turn-left-only maze.

Fig. 4.27   A maze on a torus

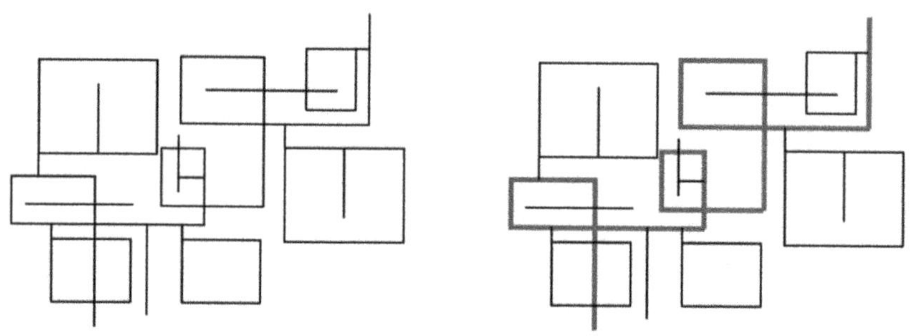

Fig. 4.28   A turn-left-only maze and its solution

Another favourite of mine are number mazes. The objective is to get from the start to the end by jumping the number of squares written in the cell you are on vertically, horizontally or diagonally. The number maze shown in Fig. 4.29 became famous. It was created and published in the *New York Journal and Advertiser* on April 24, 1898, by American puzzle maker Sam Lloyd, who called it "Back to the Klondike".

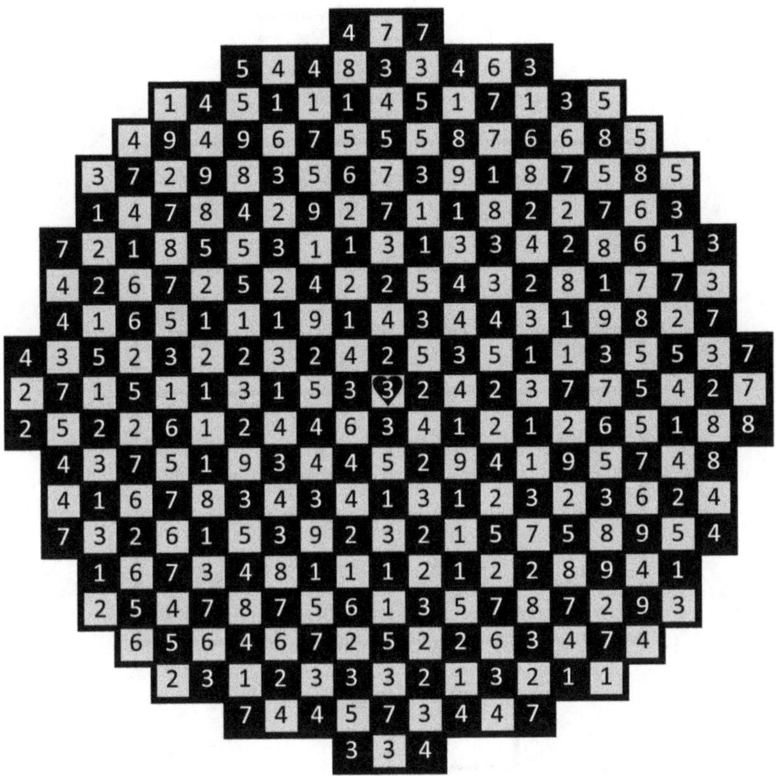

Fig. 4.29   Sam Lloyd's "Back to the Klondike"

You start at the centre of the maze marked with a heart shape and the number 3, which indicates that your first move should be 3 squares horizontally, vertically or diagonally. The object is to exit the maze, *but*, your last move has to be just one step beyond the border. This means that if you land on the number 4 in the bottom row, you cannot exit the maze because you would be overjumping the border by 3 steps. You can try and find the single solution to the maze if you like, but I warn you, it's very difficult. If you do, do not read the next line, which gives the solution using the abbreviations S for South, N for North, W for West, and E for East. SW, SW, NE, NE, NE, SW, SW, SW, SE.

Fig. 4.30   An eyeball maze

Eyeball mazes are very easy to make yet can be quite difficult to solve. They are designed so that you don't need to write to solve them. You only need, well, your eyeballs! Figure 4.30 shows an eyeball maze. You have to get from the entry point, which is the cell with the eyeballs (on the bottom row), to the exit by moving horizontally, vertically or diagonally without skipping cells, but you are only allowed to move to cells that have either the same shape or the same colour as the one you are currently on.

Possibly, the most imaginative use of mazes is to design and solve mechanical puzzles. Mechanical puzzles are physical objects, usually made of plastic, metal or wood, that you must move, arrange or solve to complete a challenge or reach a goal. It turns out that many mechanical puzzles are isomorphic to mazes, so instead of solving the puzzle, you can map it onto a maze and solve the maze. Kenneth E. Caviness, a physicist at the Southern Adventist University in Collegedale, Tennessee, has made a great Mathematica™ program where you can create and solve virtual "free the key" puzzles and their maze counterparts. The "free the key" puzzle consists of a key with teeth at different heights and a flat disc with different-sized cavities in different directions so that, to put the disc through the key — either to take it off or put it on — you have to rotate the disc so that the cavities match the teeth. If you've ever tried to solve one of these puzzles, you'll probably agree that it's not as easy as it might seem. In his program, the puzzle is mapped to a cylindrical maze in the following way. Suppose there are 16 cavities on the disc and 21 teeth on the key. Number the cavities with numbers 1 through 16, starting at any arbitrary point. Similarly, number the teeth according to their height along the key. Construct a table where the rows represent the cavities and the columns the teeth. Blacken out all the cells (i, j) in the table where the ith cavity is not large enough for the kth tooth to go through. For example, blacken out cell $(3, 17)$ if cavity number 3 is not large enough for the tooth at height 17 to go through. These are the walls of the maze. The remaining unblackened cells are the path. Solve the maze — and you've solved the puzzle, and vice versa! Figure 4.31 shows a "free the key" puzzle and its maze that I made using Kenneth's program.

By now, you might want to create your own maze. There are many websites that give step-by-step instructions on how to draw your own maze or labyrinth or how to generate them through various computer software. My favorite is Jason Roush's "Do you maze?" website.

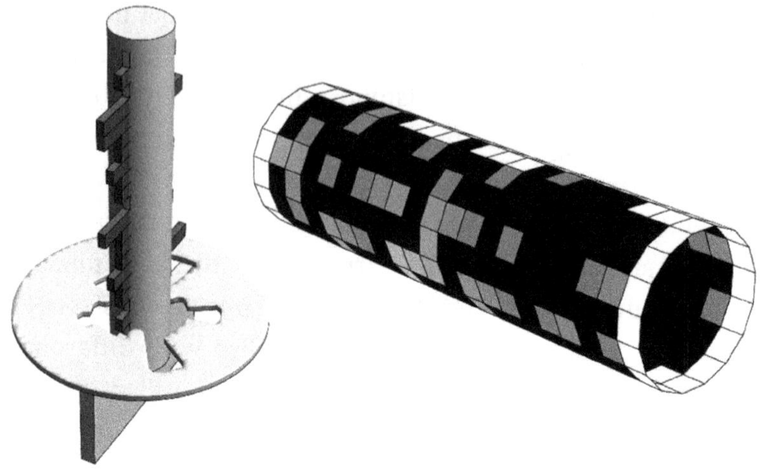

Fig. 4.31   A "free the key" puzzle and its maze

Recap

This chapter has covered an awful lot of material, and we still have barely touched the topic. We could have chosen to elaborate on the number theory behind SAT sequences, the combinatorics of counting maze solutions, or more graph theory and its connection to mazes. And we could have seen many more maze types or discussed their use in modern-day science and technology. Alas, we just don't have enough space to cram everything in. There are, however, many resources where you can learn more, and you'll find some useful references in the bibliography. We end with some of the definitions and concepts that we encountered along the way, the largest inventory in this book!

- *Maze* — a puzzle made from a network of paths. The aim is to navigate from a source point, usually the entrance, to a target point, usually either the exit or the centre. A maze can have more than one source and target point.

- *Labyrinth* — a maze with only a single path leading to its centre. It is also called a *unicursal maze*.

- *Simple maze* — a labyrinth where each level is traversed only and exactly once.

- *Alternating maze* — a labyrinth structured so that the levels are traversed clockwise and anti-clockwise, alternately.

- *Transit maze* — a labyrinth structured so that there is only one path to the centre, no other way out, and no branches.

- *SAT maze* — a simple, alternating transit labyrinth.

- *Levels of a maze* — The concentric layers that make up the labyrinth.

- *Level sequence* — The series of numbers representing the levels in the order the maze solver encounters on his journey towards the centre.

- *Proof by contradiction* — Make the opposite assumption to the given statement and show that you inevitably reach a logical fallacy.

- *Parity* — The property of being odd or even.

- *Isomorphic* — two things that look different on the outside but have the same underlying form or structure.

- *Meander* — a self-avoiding curve that crosses a straight line a number of times. If the beginning and the end of the curve are connected, the meander is called a *closed* meander; otherwise, it is an *open* meander.

- *Meander sequence* — the series of numbers that you get using the following algorithm. Assign a whole number to each part of the curve, beginning with 1, and increase it by one each time it crosses the line. Read off the numbers from left to right along the straight line. The equivalent SAT-maze sequence is obtained by adding a zero at the beginning of the Meander sequence.

- *Simply connected maze* — a maze where all the walls are connected.

- *Backtracking algorithm* — a general term for a class of algorithms frequently used by computer scientists to solve problems. Traverse a network of potential paths to solutions, make a choice at each decision junction along the way, and upon reaching an invalid solution, discard it and return to the previous decision junction to make a new, different choice.

- *Deep search algorithm* — a backtracking algorithm for searching mazes and similar structures where you initially travel deeper and deeper into the maze and then work your way back (backtrack). Also known as *Trémaux's algorithm*.

- *Dead-end algorithm* — Given a maze, find all the dead ends and "fill them in" up to the first junction.

- *Graph* — a collection of points, called vertices, connected by lines, called edges, that represent relationships or connections between the points.

- *Simple graph* — a graph with exactly one edge between any two vertices and no vertices are connected to themselves.

- *Complete graph* — a graph where all the vertices are connected to each other.

- *Disconnected graph* — a graph with detached regions.

- *Directed graph* — a graph with edges that indicate a one-way relationship between two vertices.

- *Cyclic graph* — a graph with at least one cycle.

- *Cycle graph* — a graph where the whole graph is a single cycle.

- *Acyclic graph* — a graph with no cycles.

- *Tree* — an acyclic graph where all the vertices are connected and with a unique path between any two.

- *Planar graph* — a graph whose edges do not cross and intersect only at their endpoints.

- *Degree* — the number of edges connected to a vertex in a graph.

- *Eulerian path* — a route that visits every edge in a graph exactly once. If the path begins and ends at the same vertex, it is called an *Eulerian cycle*, and if it begins at one vertex and ends at another, it is called an *Eulerian trail*.

- *Euler's theorem on graphs* — an Eulerian cycle exists on connected graphs with all-even vertex degrees, and an Eulerian trail exists on those with two odd vertex degrees.

- *Hamiltonian path* — a route that visits every vertex in a graph exactly once. Hamiltonian cycles and trails are defined similarly to their Eulerian counterparts with respect to their beginning and endpoints.

- *Knight's tour* — a path that visits every square on a board exactly once while moving like a knight.

- *Platonic solids* — The five solids with the properties that all the angles and all the edge lengths are equal and the same number of faces meet at each vertex. They are the tetrahedron, the cube, the octahedron, the dodecahedron, and the icosahedron.

- *Euler's formula for spherical polyhedra* — the sum of the number of faces and the number of vertices minus the number of faces in a spherical polyhedron is equal to 2.

- *Euler's characteristic* — a number that describes the topology of a polyhedra.

- *Perfect maze* — also called a *simply connected maze*, is a maze that has no loops or circuits, and there is exactly one path from entry to exit.

- *Braid maze* — also called a *purely multiply-connected maze*, is a maze that doesn't have dead ends.

- *Arrow maze* — a directed graph where, when you land on an arrow, you can move to any of the cells in either the row, column, or diagonal that it points to.

- *Gamma mazes* — mazes whose basic cells are rectangular and whose paths meet at right angles.

- *Delta mazes* — mazes whose basic cells are equilateral triangles and whose paths meet at multiples of 120° angles.

- *Sigma mazes* — mazes whose basic cells are regular hexagons and whose paths meet at multiples of 120° angles.

- *Upsilon mazes* — mazes whose basic cells are a combination of squares and octagons.

- *Theta mazes* — mazes whose cells are arranged in concentric circles.

- *Crack mazes* — mazes whose cells are arranged with no consistent tessellation.

- *Logic maze* — a maze with additional rules.

- *Number maze* — a maze where you have to get from start to end by jumping the number of squares written in the cell you are on vertically, horizontally or diagonally.

- *Eyeball maze* — a maze where you have to get from the cell with the eyeballs to the exit by moving horizontally, vertically or diagonally without skipping cells, moving only to cells that have either the same shape or the same colour as the one you are currently on.

Challenge Yourself!

(1) Which of the following sequences is not a SAT maze, and why?

   (a) 0, 3, 2, 1, 4, 7, 6, 5, 8

   (b) 0, 1, 2, 3, 4, 5

   (c) 0, 7, 8, 5, 6, 2, 9, 1, 10, 3, 4, 11, 12

   (d) 0, 3, 6, 7, 4, 5, 8, 1, 2, 9, 10

   (e) 0, 1, 8, 5, 6, 3, 2, 7, 4

(2) Draw a meander and a labyrinth for the SAT sequence: 0, 1, 2, 5, 4, 3, 6, 7, 8

(3) Name the three graphs in Fig. 4.32.

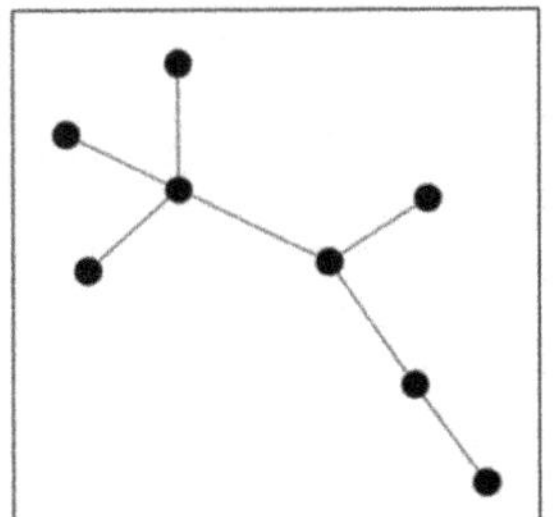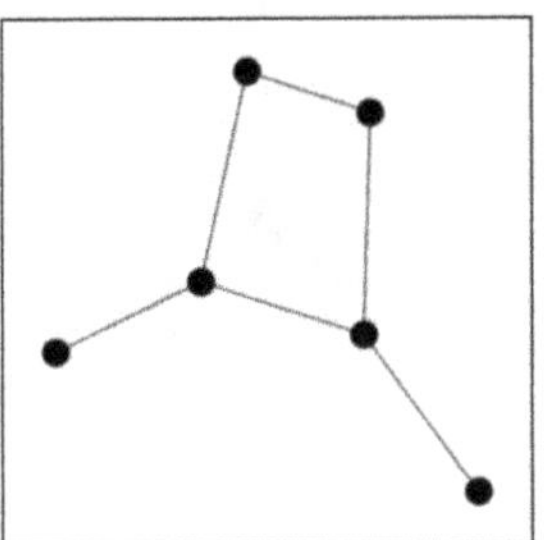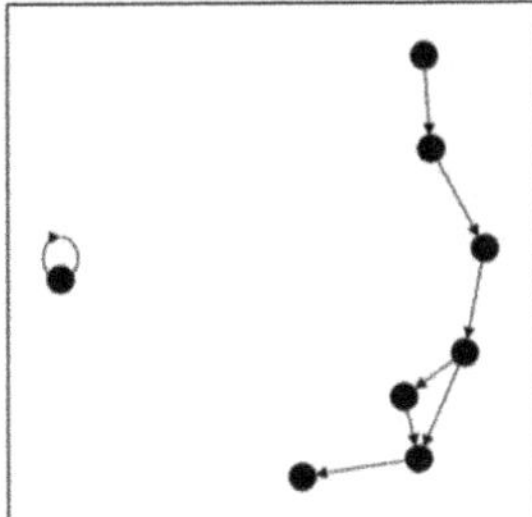

Fig. 4.32   Name the three graphs in this figure

(4) Which of the diagrams in Fig. 4.33 can be drawn without taking your pencil off the paper?

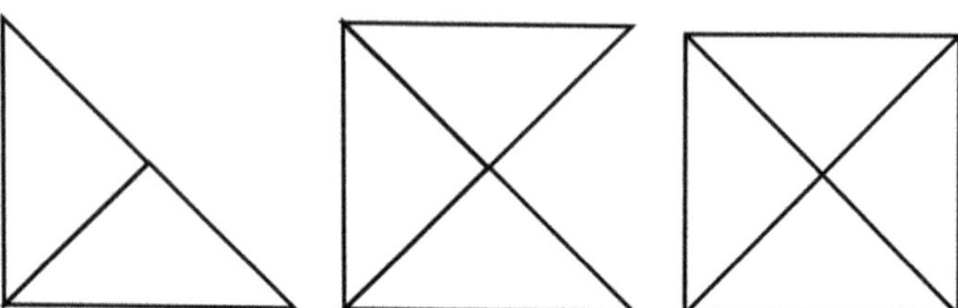

Fig. 4.33   Which of these can be drawn without taking your pencil off the paper?

(5) Figure 4.34 shows a connected, cyclic graph. Transform it into a maze by changing the vertices into junctions and the edges between vertices into the corresponding paths between the junctions.

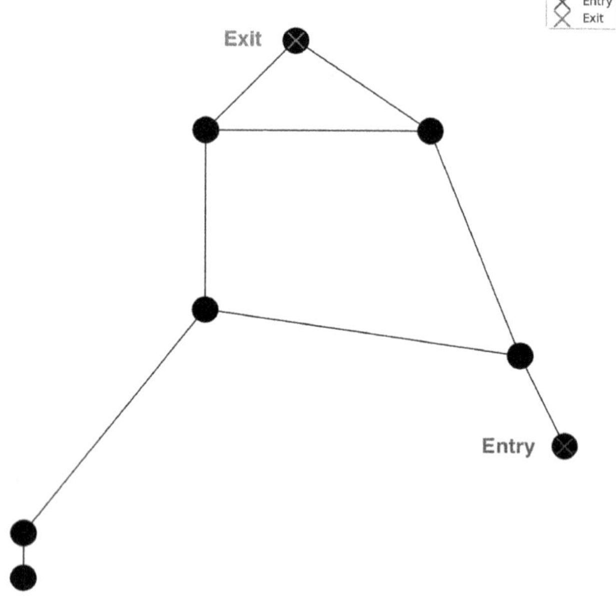

Fig. 4.34   A connected, cyclic graph waiting to be transformed into a maze!

(6) Solve the classic maze in Fig. 4.35.

Fig. 4.35   A classic maze

(7) Solve the arrow maze in Fig. 4.36.

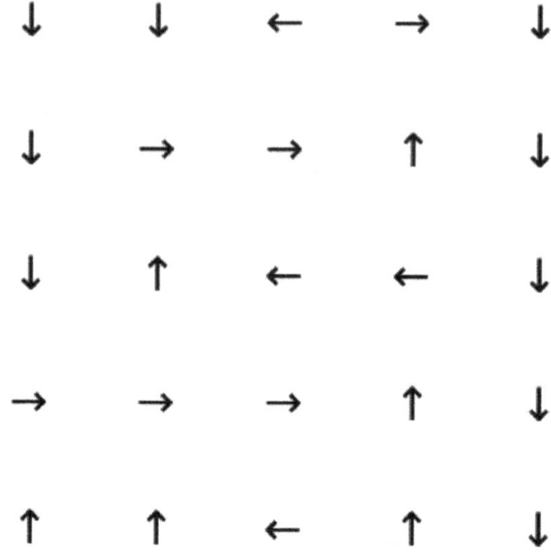

Fig. 4.36   An arrow maze

(8) Solve the number maze in Fig. 4.37.

START 2	3	3	4	3
2	1	2	2	2
4	1	2	2	3
1	1	2	2	4
3	3	3	3	END

Fig. 4.37   A number maze

(9) Elizabeth Carpenter, founder of Mazeology Design, is a New York City-based multi-award-winning maze-maker, game designer, illustrator, and author. She has developed a unique maze design form, where she turns line art illustrations into mazes, which have been published in books and magazines, and provides mazes to corporate clients. She very graciously gave me permission to reproduce a maze (Fig. 4.38) she designed in honour of the legendary Martin Gardner, one of the most proficient and popular math and science writers of the 20th century. Apart from it being a great maze, she illustrated the maze with themes that Martin Gardner would enjoy — recreational math, puzzles, math art, Alice in Wonderland, and lots of other stuff. Solve the maze and find as many Martin Gardner themes as you can.

Fig. 4.38   Elizabeth Carpenter's Martin Gardner maze

Solutions

(1) All sequences are SAT sequences except for (c) and (e). Not all the numbers in sequence (c) are alternating odd and even numbers. Sequence (e) does not end with the last level, 8.

(2) Figure 4.39 shows the meander, and Fig. 4.40 shows the labyrinth.

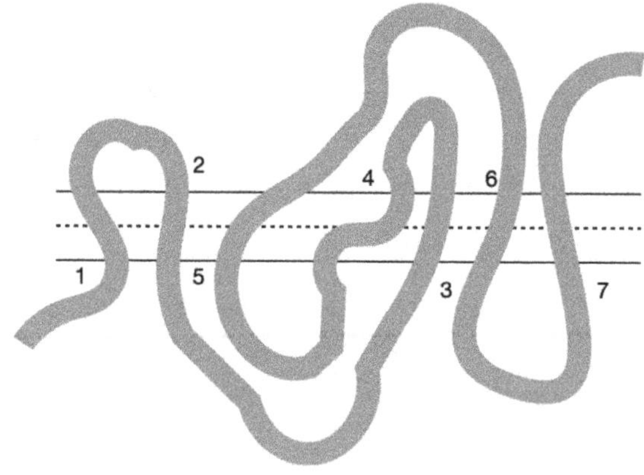

Fig. 4.39   Meander for SAT sequence: 0, 1, 2, 5, 4, 3, 6, 7, 8

Fig. 4.40   Labyrinth for SAT sequence: 0, 1, 2, 5, 4, 3, 6, 7, 8

(3) From left to right: a simple, connected, undirected, acyclic graph; a connected, cyclic, undirected graph; and an unconnected, directed graph with a loop.

(4) The leftmost and centre images both have two odd vertex indexes, and the rest are even, so they can be drawn in one go. The rightmost image has four odd vertices, so it cannot.

(5) The maze that I drew is shown in Fig. 4.41. Yours might be slightly different, but the number of paths and vertices and their connectivity should be the same.

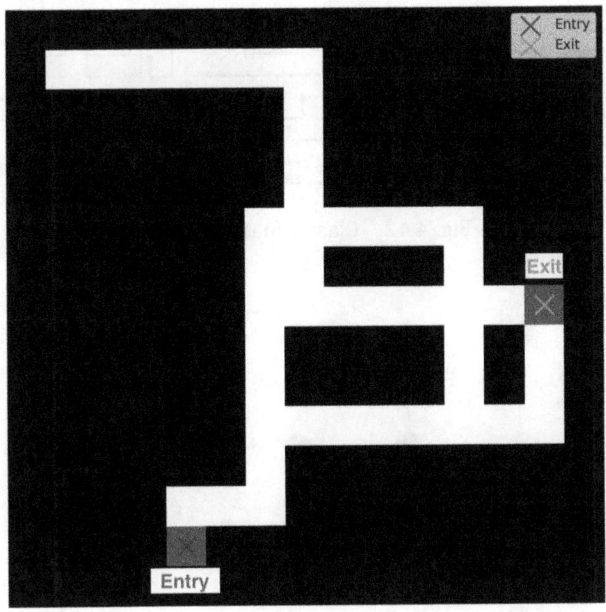

Fig. 4.41   Maze corresponding to the graph in Fig. 4.34

(6)

Fig. 4.42   Classic maze solution

(7)

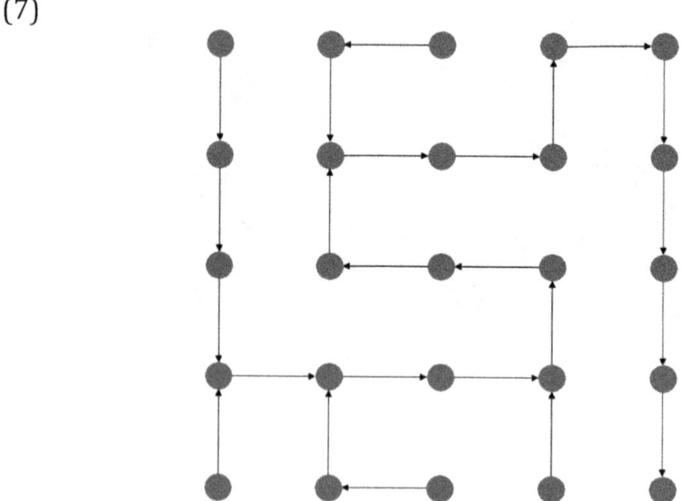

Fig. 4.43   Arrow maze solution

(8) The solution is: right-down-left-right-down-right.

(9) Following the solution are Elizabeth Carpenter's notes about it.

Fig. 4.44   Elizabeth Carpenter's Martin Gardner maze solution

1. This illustration is a MAZE. To find the maze's START and FINISH, solve the rebus puzzles at the top right corner and the bottom left corner.

2. To see the Mad Hatter sip his tea, hold the illustration in front of you and keep your eyes on the tiny "x". Slowly bring the drawing to your face until the tip of your nose touches the "x".

3. Above the "x" is a "Ghost Triangle".

4. Martin Gardner was fond of card tricks. These cards represent prime numbers.

5. The portrait is framed by an impossible crate.

6. The Möbius strip is a homage to the M.C. Escher's ant Möbius strip.

7. To the portrait's left is an image of Alice stepping into the mirror.

8. Below Alice is one of Martin's matchstick puzzles: Ten matches are sketched as an equation in Roman numerals. But the equation is incorrect. Can you correct the equation without changing any of the matches, adding any new matches, or removing any of the matches?

9. At the centre of the page, below the portrait's chin is Martin's "Dime-and-Penny Switcheroo". There are two different types of moves you can make: You can either slide a coin into an empty space that is next to it or you can jump a coin over another coin (as in checkers), provided that the coin lands in an empty space. The goal is to see if you can get the dimes and pennies to switch places in exactly eight moves. According to Martin, if you can solve the Switcheroo in under 5 minutes, you are a genius!

10. On the left side of the portrait's sweater is a sketch of M.C. Escher's mathematical mosiac *Geese*, which was featured on the cover of *Scientific American*, April 1961.

11. Below *Geese* is one of Martin's toothpick puzzles: In this sketch 16 toothpicks make 5 squares. Can you change the positions of two toothpicks to make four squares, all the same size, with no toothpicks left over?

12. On the right side of the portrait's sweater is a sketch of the Infinite Star Pattern that was featured on the cover of *Scientific American*, January 1977.

13. Lastly, the nameplate is a wonderful Ambigram that was designed by Scott Kim that spells out Martin's alter-ego. Turn the illustration over to reveal the true name of the portrait in the maze.

Acknowledgements

I would like to thank Dick Esterle, Sam Freund, and Peter Caesar for sharing their enthusiasm, suggestions and ideas while I was creating the illustration for this maze. Special thanks to Sam for also testing this maze.

A very special thank you to Scott Kim for allowing me to use his fabulous Ambigram of Martin's name for the nameplate. Ambigram lettering art is ©Scott Kim, scottkim.com.

Bibliography and Further Reading

Caviness, K. E. (2011). Maze for Free the Key Puzzle. *The Mathematica Journal*, dx.doi.org/doi:10.3888/tmj.13–1

Dawson, R. (2021). Maze generator. *Maze Generator*. Retrieved August 22, 2024, from https://codebox.net/pages/maze-generator.

Kappraff, J., Radović, L., and Jablan, S. (2016). Meanders, knots, labyrinths and mazes. *Journal of Knot Theory and Its Ramifications*, **25**, 9, 1641009. https://doi.org/10.1142/s0218216516410091

Matthews, W. H. (1922). *Mazes and Labyrinths: A General Account of Their History and Development*. Longmans, Green and Co. London. Project Gutenberg. Retrieved August 18, 2024, from https://www.gutenberg.org/ebooks/46238.

Niemczyk, R. and Zawiślak, S. (2020). Review of maze solving algorithms for 2D maze and their visualisation. In *Engineer of the XXI Century Cham*, (eds.) S. Zawiślak and J. Rysiński.

O'Connor, J. J. and Robertson, E. F. (1999). William Rowan Hamilton. *MacTutor Index.* School of Mathematics and Statistics, University of St Andrews, Scotland. Retrieved August 29, 2024, from https://mathshistory.st-andrews.ac.uk/Biographies/Hamilton/.

Phillips, T. (2024). Through mazes to mathematics. *Tony Phillips Homepage.* Retrieved August 22, 2024, from https://www.math.stonybrook.edu/~tony/mazes/index.html.

Pullen, W. (2022). Maze classification. *Think Labyrinth!* Retrieved August 22, 2024, from https://www.astrolog.org/labyrnth/algrithm.htm.

Roush, J. (2019). Maze collections. *Do You Maze?* https://www.doyoumaze.com/maze-collections

Timperi, K. G., LaValle, A. J., and LaValle, S. M. (2024). *Universal Plans: One Action Sequence to Solve Them All!* https://arxiv.org/abs/2407.02090

Wikipedia contributors. (2024, July 23). Maze-solving algorithm. *Wikipedia, The Free Encyclopedia.* Retrieved August 26, 2024, from https://en.wikipedia.org/w/index.php?title=Maze-solving_algorithm&oldid=1236152869.

Wikipedia contributors. (2024, July 25). Platonic solid. *Wikipedia, The Free Encyclopedia.* Retrieved August 29, 2024, from https://en.wikipedia.org/w/index.php?title=Platonic_solid&oldid=1236642411.

Wikipedia contributors. (2024, August 5). Meander (art). *Wikipedia, The Free Encyclopedia.* Retrieved August 22, 2024, from https://en.wikipedia.org/w/index.php?title=Meander_(art)&oldid=1238747495.

Wikipedia contributors. (2024, August 20). Hamiltonian path problem. *Wikipedia, The Free Encyclopedia.* Retrieved August 29, 2024, from https://en.wikipedia.org/w/index.php?title=Hamiltonian_path_problem&oldid=1241362622.

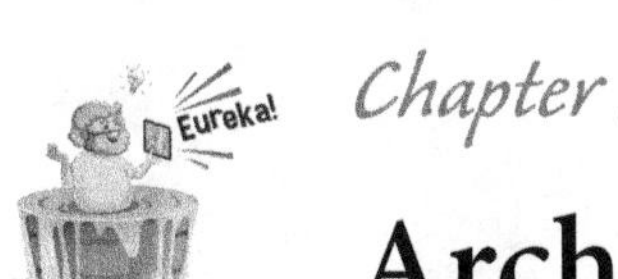

Chapter 5

Archimedes' Stomach

Introduction

Archimedes' Stomach, better known as Archimedes' Ostomachion, is arguably considered "the world's oldest puzzle". It is a classic geometry challenge in line with the main mathematical pursuit of ancient Greek mathematicians. Geometry constructions and dissections are, of course, difficult to do by reading a book, so I urge you to print and cut out the Ostomachion from Wikipedia or other online sources.

The Puzzle

Rearrange the following pieces into a square (Fig. 5.1). Find at least two solutions.

Fig. 5.1   Archimedes' Ostomachion puzzle pieces

Where to Start?

First, you need to cut out the pieces. Please do not ruin the book! Just make a copy on a piece of paper or cardboard. To solve the puzzle, trial and error is your best bet, at least until we learn some strategies later in the chapter.

Solving the Puzzle

It was only in 2003 that Bill Cutler, a mathematician and puzzle maker, proved that there are 536 distinct solutions to Archimedes' Ostomachion. Here they are (Fig. 5.2), beautifully recreated in one frame by Japanese software engineer Tetsunori Nakayama.

Fig. 5.2   The 536 solutions to Archimedes' Ostomachion puzzle

Of course, you are going to need a microscope to see an individual solution, but you can go online to Tetsunori Nakayama's open processing gallery and zoom in to the whole array of solutions online (https://openprocessing.org/sketch/1041477) or view individually, randomly generated solutions (https://openprocessing.org/sketch/1048183). Figure 5.3 shows four individual solutions from the gallery.

Fig. 5.3   Four random solutions to Archimedes' Ostomachion puzzle

The History of the Puzzle and its Inventor

Archimedes is one of the world's most famous Greek mathematicians. Born in 267 B.C. in Syracuse, Sicily, he was far more than a mathematician — he dabbled in almost anything from astronomy to engineering. You can see a beautiful rendering of Archimedes in *Portrait of a Scholar*, painted by Domenico Fetti in 1620 and on display at Gemäldegalerie Alte Meister in Dresden, Germany (Fig. 5.4). Note the globe in his right hand, the circles and geometrical shapes on the paper, and the ruler and compass on the table.

Archimedes regularly communicated with scholars in Alexandria, Egypt, the location of the Ivy League schools of those days, and is rumoured to have visited them at least once. It is said that it is on

Fig. 5.4   *Portrait of a Scholar*, Domenico Fetti, 1620

one of these visits that he invented what is now known as Archimedes' screw, a large screw-shaped contraption used to lift water from a lower to a higher level. This giant screw has a spiral trough that, when turned, causes the water to flow up through it and then out of an opening near the top. Used throughout history, along the Nile and other places around the world, Archimedes' screw is still used today for irrigation (Fig. 5.5), water supply, and sewage treatment. You can see a working model of the screw in many places, including the Clore Garden of Science, an open-air science museum at the Weizmann Institute of Science in Rehovot, Israel.

It is impossible to even begin to summarise Archimedes' biography and huge contributions to mankind. His accomplishments include inventing "war machines" that were used in battle and built

Fig. 5.5   An Egyptian working on the field using an irrigation pump — Archimedes' screw

on physics principles. Here is some of what Plutarch has to say about Archimedes in his biography "The Life of Marcellus", which appears in "The Parallel Lives":

"And yet even Archimedes, who was a kinsman and friend of King Hiero, wrote to him that with any given force it was possible to move any given weight; and emboldened, as we are told, by the strength of his demonstration, he declared that, if there were another world, and he could go to it, he could move this. Hiero was astonished, and begged him to put his proposition into execution, and show him some great weight moved

by a slight force. Archimedes therefore fixed upon a three-masted merchantman of the royal fleet, which had been dragged ashore by the great labours of many men, and after putting on board many passengers and the customary freight, he seated himself at a distance from her, and without any great effort, but quietly setting in motion with his hand a system of compound pulleys, drew her towards him smoothly and evenly, as though she were gliding through the water. Amazed at this, then, and comprehending the power of his art, the king persuaded Archimedes to prepare for him offensive and defensive engines to be used in every kind of siege warfare... When, therefore, the Romans assaulted them by sea and land, the Syracusans were stricken dumb with terror; they thought that nothing could withstand so furious an onset by such forces. But Archimedes began to ply his engines, and shot against the land forces of the assailants all sorts of missiles and immense masses of stones, which came down with incredible din and speed; nothing whatever could ward off their weight, but they knocked down in heaps those who stood in their way, and threw their ranks into confusion. At the same time huge beams were suddenly projected over the ships from the walls, which sank some of them with great weights plunging down from on high; others were seized at the prow by iron claws, or beaks like the beaks of cranes, drawn straight up into the air, and then plunged stern foremost into the depths, or were turned round and round by means of enginery within the city, and dashed upon the steep cliffs that jutted out beneath the wall of the city, with great destruction of the fighting men on board, who perished in the wrecks..."

Obviously, there is much exaggeration in Plutarch's report, but still, we can feel the immense respect the Syracusans had for Archimedes and his inventions. Note Plutarch's mention of the well-known expression attributed to Archimedes, who was obsessed with levers and pulleys: "Give me a lever long enough and a pivot on which to place it, and I shall move the world."

Archimedes was always looking for new ways to explain the physical world. Astronomy, optics, physics, hydrostatics — you name it, Archimedes is sure to have made some significant contributions, and I urge you to read some of the reference material at the end of the chapter to grasp the extraordinary breadth of topics that Archimedes was involved in. He was a true "absent-minded" scientist, as witnessed by his alleged cry of "Eureka" when he discovered one of the laws governing floating bodies while taking a bath (historians dispute which law he actually discovered in the bath). Indeed, the bath was one of his favourite places for doing math, at least according to Plutarch, who wrote:

> *"And therefore we may not disbelieve the stories told about him, how, under the lasting charm of some familiar and domestic Siren, he forgot even his food and neglected the care of his person; and how, when he was dragged by main force, as he often was, to the place for bathing and anointing his body, he would trace geometrical figures in the ashes, and draw lines with his finger in the oil with which his body was anointed, being possessed by a great delight, and in very truth a captive of the Muses. And although he made many excellent discoveries, he is said to have asked his kinsmen and friends to place over the grave where he should be buried a cylinder enclosing a sphere, with an inscription giving the proportion by which the containing solid exceeds the contained."*

In this paragraph, Plutarch refers to Archimedes' true love: math, and particularly geometry. Archimedes is credited with many discoveries in geometry, including formulas for calculating the areas of a triangle and a circle and the volumes of spheres and cylinders. Indeed, upon his request, his gravestone was engraved with pictures of a sphere inscribed within a cylinder, illustrating his discovery that the ratio of their volumes is two to three. Many geometric objects are named after him: Archimedean Triangles, Archimedean Circles, and the Archimedean Solids, to name but a few. His calculation of pi (3.141...) was ingenious. Archimedes knew that pi was an irrational number — a real number that cannot be expressed as a ratio of two integers or as a repeating or finite decimal — but he wanted to get the most accurate value that he could. The way this was done was to inscribe a circle with a 96-sided regular polygon and circumscribe it with another polygon (Fig. 5.6). This method had been used before, but only with smaller polygons like squares and hexagons.

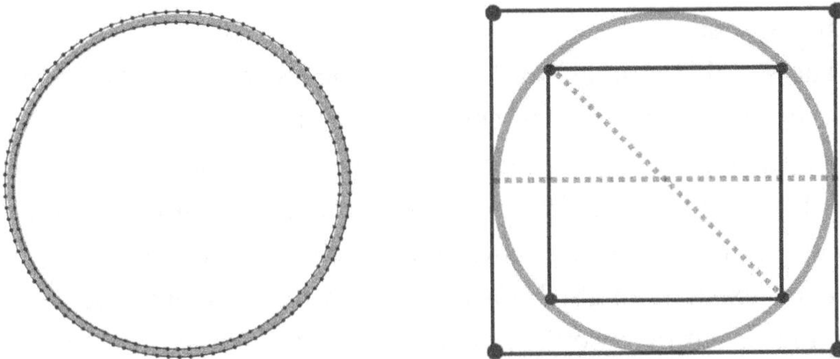

Fig. 5.6 A regular 96-gon is inscribed in a circle circumscribed by a regular 96-gon (left); a square is inscribed in a circle circumscribed by a square (right). The dotted lines show the circle's diameter as the midsection of the outer square and the diagonal of the inner square

The diameter of the circle can be calculated exactly by applying elementary geometry to the diagonals and midsections of the inner and outer polygons. The circumference of the circle is given two approximations: one by the circumferences of the outer polygon and the other by the circumference of the inner polygon. Pi is approximated to be a value between an upper boundary given by dividing the circumference of the outer polygon by the diagonal and a lower value given by the same approximation using the circumference of the inner polygon. Archimedes' value of pi was:

$$\frac{223}{71} < \pi < \frac{22}{7}$$

This is perhaps the origin of the common mistake some people still make of thinking that pi equals $3\frac{1}{7}$.

How do we know so much about Archimedes? For one thing, he wrote many books and treatises. Some survived through the many centuries since his death; others were lost. *On the Sphere and the Cylinder*, *On Conoids and Spheroids*, *On Floating Bodies*, and *On Spirals* are but a few of his surviving works on geometry alone. Archimedes also wrote *Ostomachion*, a treatise describing our puzzle and its geometrical aspects.

The word "Ostomachion", which appears in a translation of the treatise written by the 4th-century Roman poet Decimius Magnus Ausonius, has been translated by some researchers as "challenge with bone pieces". This does seem more appropriate than the puzzle's other name, Stomachion, which, in Ancient Greek, does not even exist! It might be a corruption of the word "stómachos", meaning throat or oesophagus or possibly the entry to the stomach. However, it gave me a better title for this book, so that is the name I am sticking to — at least for the cover!

The Ostomachion is considered the European counterpart of the Chinese tangram (Fig. 5.7), a similar dissection puzzle made of seven different parts. Given that the tangram is much more popular and well-known than the Ostomachion, it is hard to believe that its first recorded mention is 1795 A.D. The Ostomachion, on the other hand, is far older.

Fig. 5.7   A tangram

Although the Ostomachion was mentioned and discussed in a few surviving translated works, Archimedes' original text was thought to be lost, along with another text on physics, until a treatise known as *Archimedes Palimpsest* surfaced just over 100 years ago.

The palimpsest, which is now on display at Walter's Art Museum in Baltimore, is a collection of parchments from several authors, including Archimedes, compiled into one codex (the predecessor of a book) by the Byzantine Greek architect, mathematician, and physicist Isidore of Miletus in the sixth century A.D., who taught at the university in Constantinople — modern-day Istanbul, Turkey. The two Archimedean texts in the document were his treatises, *The Method of Mechanical Theorems* and *Ostomachion*. The original has never been found, but luckily, an anonymous Byzantine Greek author made a copy of the original codex in the tenth century. Following the crusades in the early 13th century, the palimpsest was taken first to Jerusalem and then to Mar Saba, a Greek Orthodox monastery nearby in the Judean desert. Mar Saba is a spectacular sight and an active monastery that you can still visit today. In those days, recycling was the mode, and the practice was to unbind unwanted codexes, wash them, and reuse them for other purposes. This is exactly what happened to the palimpsest in either Jerusalem or Mar Saba — the parchments were unbound, washed, and reassembled along with other parchments, and once they were more or less clean, they were re-used to produce a 177-page Christian prayerbook (Fig. 5.8).

In the 17th or 18th century, this prayerbook made its way back to the Greek Orthodox library in Constantinople, where it gained the attention of three scholars: Bible scholar Constantin von Tischendorf, who first noticed that the faint Greek mathematics underlying the text in 1840 and stole a page of the text for a more scrupulous study, Greek librarian Papadopoulos-Kerameus, who meticulously catalogued the library and included a transliteration of some of the underlying text in the palimpsest, and Danish historian Johan Ludvig Heiberg, the world's leading authority on Archimedes who, in 1906, visited Constantinople, photographed the prayer book and immediately understood its immense value. But the saga of recovering Archimedes' lost texts had only just begun.

Heiberg's work on the palimpsest was published in the early 20th century and was translated into English, but the actual document was again lost during the 1920s Turkish–Greek wars. It resurfaced in a Christie's auction in 1998, proposed by the daughter of a French businessman, Marie Louis Sirieix, who claimed that her father bought the palimpsest from a monk in Turkey, but some suspect that he actually stole it. Indeed, the Greek Orthodox authorities tried to stop the auction legally, disputing Sirieix's ownership, but they lost the case due to the long time that had passed, and the codex finally came to its current resting place in Walters Art Museum in Baltimore, on loan by the anonymous person who won the auction and bought the palimpsest, allegedly, Amazon's CEO Jeff Bezos. The century that passed while in the hands of the Sirieixs did not do well for the palimpsest. It was kept under very poor conditions, and forgers even added gold leaf portraits to four pages of the book to make it look as if it was decorated with medieval evangelical artwork and, hence, more valuable.

Fig. 5.8   A page from *Archimedes Palimpsest* where you can see the
text of the prayer book and the faint underlying original text

It seemed that nothing from the ancient text could be salvaged, but science came to the rescue. Throughout the last twenty years, researchers from Rochester Institute of Technology and from Stanford University used modern imaging techniques, bombarding the delicate codex with X-ray, UV, IR, and other wavelengths of light, together with state-of-the-art computer imaging software, and managed to reconstruct the fullest version of the text to date.

Returning to the tangram, one of the reasons it became so popular was because you can create many shapes with the set — and these shapes became popular puzzles in their own right. You can go online and find many such challenges. Figure 5.9 shows two such puzzles (solved).

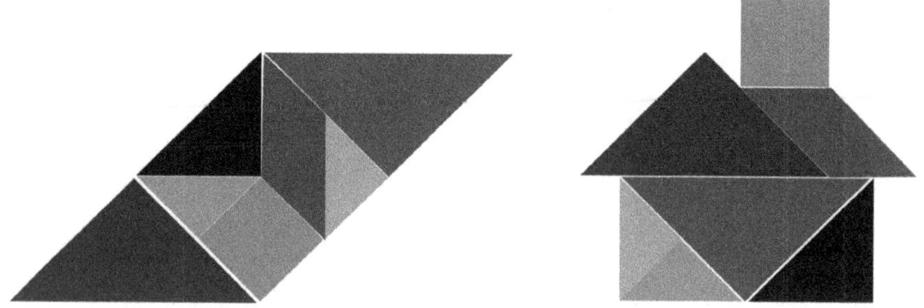

Fig. 5.9   Two solved tangram puzzles

Even in this respect, the Ostomachion beats the tangram. As early as fourth century A.D., the poet Ausonius wrote in his book *Liber XVII Cento Nuptialis*:

"So that you may say it is like the puzzle which the Greeks have called ostomachia. There you have little pieces of bone, fourteen in number and representing geometrical figures. For some are equilateral triangles, some with sides of various lengths, some symmetrical, some with right angles, some with oblique: the same people call them isosceles or equal-sided triangles, and

also right-angled and scalene. By fitting these pieces together in various ways, pictures of countless objects are produced: a monstrous elephant, a brutal boar, a goose in flight, and a gladiator in armour, a huntsman crouching down, and a dog barking — even a tower and a tankard and numberless other things of this sort, whose variety depends upon the skill of the player."

Figure 5.10 shows some of the shapes that resemble Ausonius' puzzles.

Fig. 5.10   Ostomachion puzzles

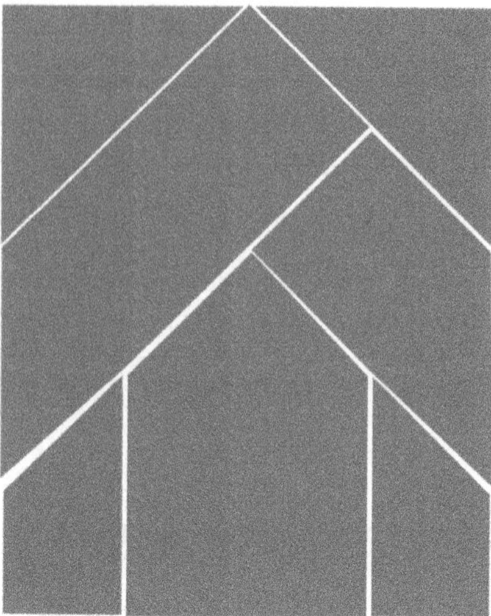

Fig. 5.11   Hanamaya's "Lucky Puzzle"

We should perhaps mention that tangrams and Ostomachions were not the only shape-filling puzzles around. Many variations were, and still are, produced by puzzle companies and popularised in various places. Japanese puzzle company Hanamaya popularised their own version in 1935 and called it "Lucky Puzzle" (Fig. 5.11). Lucky Puzzle instantly became a hit and outshined the tangram.

The mathematical study of the puzzle is far more recent and only really began with Bill Cutler's systematic analysis. Bill Cutler became interested in solving the problem after being approached by puzzlist Joe Marasco and puzzle designer Kate Jone. Cutler found the 536 solutions by enumerating all possible configurations on a computer. Unknown to Cutler, Reviel Metz, Professor of Classics at Stanford University, who was one of the two main researchers to study Archimedes' palimpsest — the other was Will Noel, the curator of manuscripts and rare books at the Walters Art Museum in Baltimore — independently approached Persi Diaconis, a Stanford mathematician, with the same task: Find all possible solutions. Diaconis collaborated with researchers from San Diego, United

States (US), Fan Chung and Ron Graham, and the three solved the problem mathematically using combinatorics — the branch in math that deals with counting things. This whole cohort of people arrived at the same answer — Bill Cutler's 536 solutions. Actually, Diaconis and co-workers got a different number, 17,152, but this is because they counted the four rotations of the puzzle, the two reflections, and the four pair exchanges of the two sets of congruent triangles as separate solutions: $536 \times 4 \times 2 \times 4 = 17152$. Figure 5.12 shows 32 solutions that Bill Cutler counted as the same solution, but the Diaconis group counted as different solutions. Can you see why Bill Cutler counted all of these as the same solution?

Persi Diaconis, Fan Chung, and Ron Graham went one step further in their investigation. They first noticed that in all the 536 solutions, three pairs of adjacent pieces always stuck together. Combining each of these pairs into one piece, they reduced the number of pieces in the puzzle to 11 and the number of solutions to 268. They called the new puzzle "Stomach" and started investigating it. They proved that all but two of the 268 solutions could be found by what they called a "simple move", meaning that you could create a new solution from a given one by flipping or rotating a symmetric region of pieces within the puzzle. Figure 5.13 shows two unique Stomach solutions that are connected by a simple move.

More recent work on the Ostomachion was initiated by Kate Jones, a remarkable puzzle researcher and designer and the co-founder and CEO of Kadon Enterprises. Kate was the main drive behind the 2003 mathematical investigations of the puzzle, culminating in the two independent 536 solution calculations. Following her passion for puzzle design, she created her own version of the Ostomachion (Fig. 5.14) by adding three colours to the pieces (the original puzzle was monochromatic). Counting different coloured solutions as separate solutions increased the number of possible solutions to $536 \times 2 = 1072$. This is because she painted two identical triangles in the Ostomachion set with different colours.

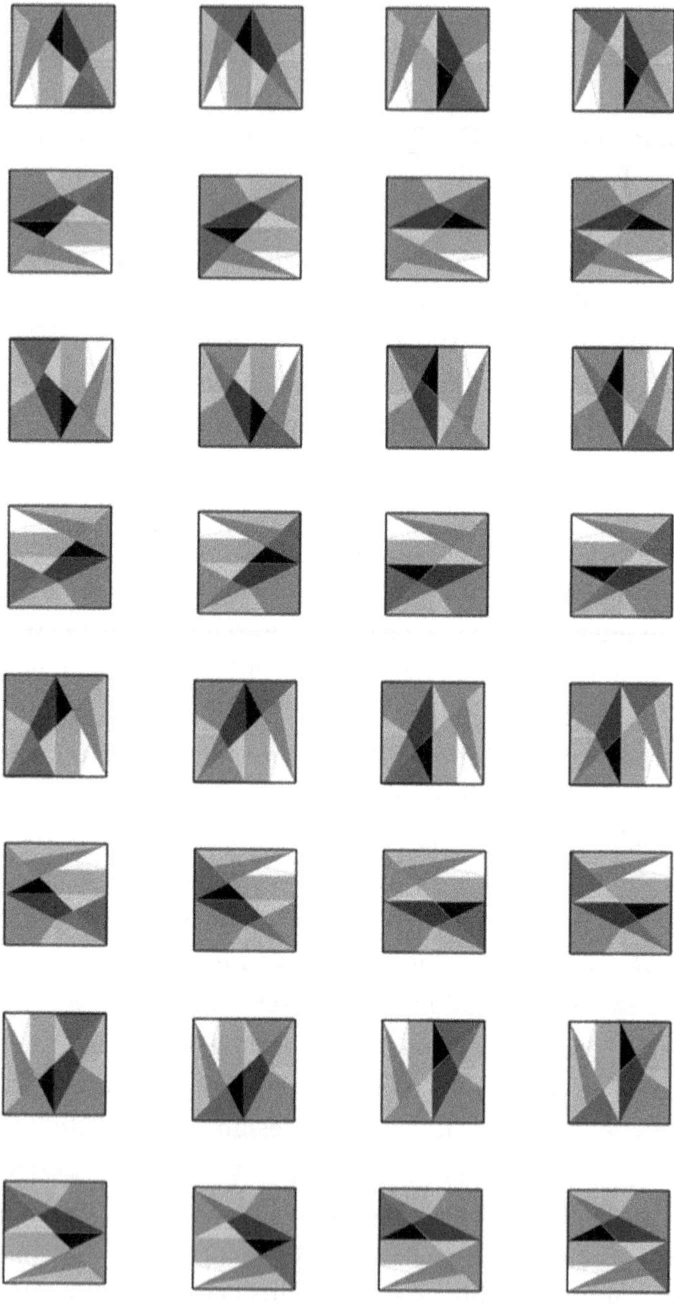

Fig. 5.12   Ostomachion solutions that Bill Cutler counted all as one

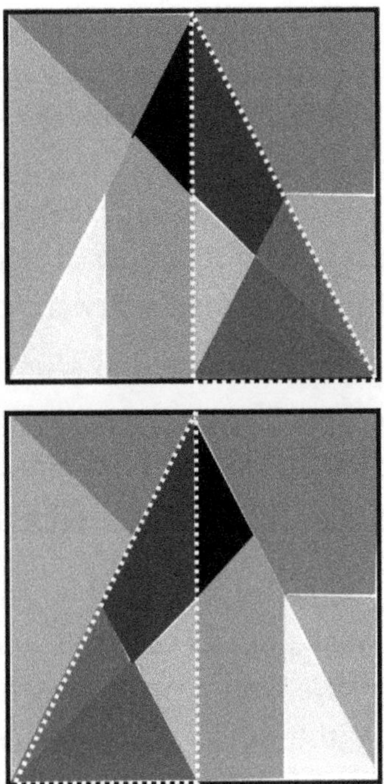

Fig. 5.13 A solution of "Stomach" (top) and another solution (bottom) generated by reflecting the pieces in the centre triangle across the midline of the square (and centre triangle)

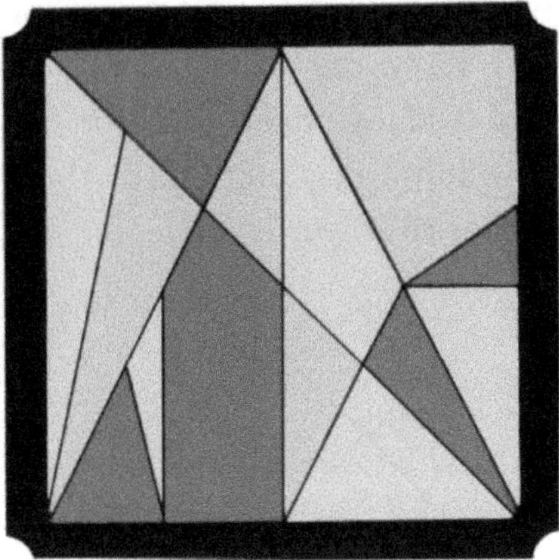

Fig. 5.14 Kate Jones' tricolour Stomachion Puzzle

Colouring does not affect rotations or reflections. It is interesting to note that Kate did not paint the pieces arbitrarily. She chose to paint them in such a way that pieces with the same colour touch at most at one point, and the total area of the pieces of each colour is the same. The tricolour Stomachion puzzle was extensively studied by Kate, Alex Streif, a research specialist at Kadon Enterprises, and Joe Marasco, a physicist, engineer, software designer, and puzzle enthusiast.

Generalisations of the Puzzle and Connections to Other Math Areas

There is so much math that can be done with tangrams, Ostomachions, and similar puzzles that it is hard to know where to start and where to stop! We have already seen how combinatorics — the math of "counting things" — and symmetry interplay while calculating the number of solutions. Then, there are the strategies used to solve the pattern puzzles. You might be thinking — strategy hunting is not math. But a lot of the time, this is *exactly* what a mathematician does. A mathematician (and computer scientist) is always on the lookout for new methods to solve problems. What strategies can be used for tangrams and Ostomachion puzzles, apart from trial and error? First, the shape of the puzzles can help us reduce the number of trials we need to make. Suppose we have the following two Ostomachion puzzles (Figs. 5.15 and 5.16). Which of the two puzzles is easier to solve?

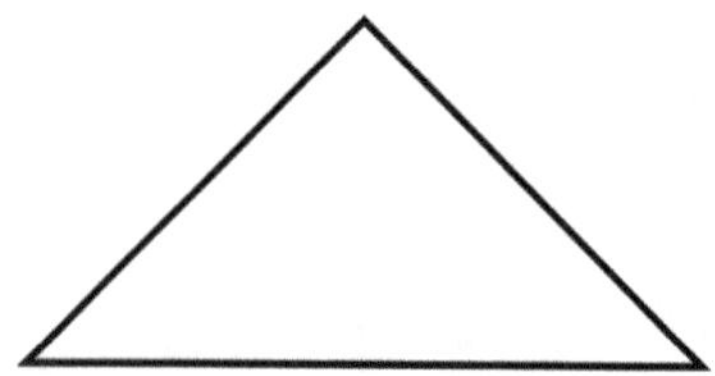

Fig. 5.15   Triangle Ostomachion puzzle

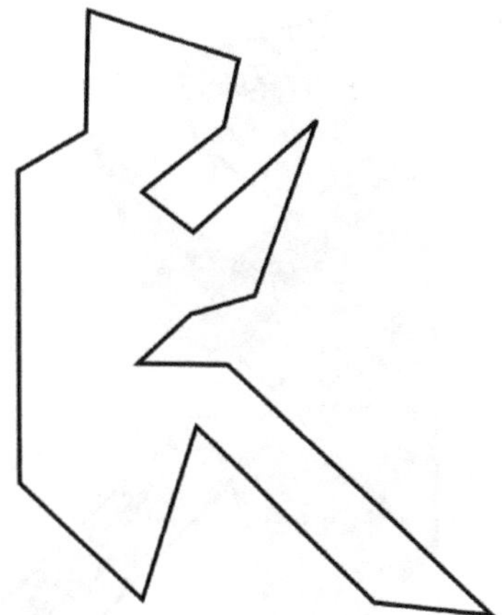

Fig. 5.16   Praying-man Ostomachion puzzle

The rule of thumb is — the more symmetry the puzzle has, the harder it is to solve. That is why the triangle is slightly easier than the square, but the praying man is much easier. The reason is the many points of symmetry breaking that hint as to which pieces need to be placed where. The man's head, arms, and legs each have very limited options for which piece goes where. Figures 5.17 and 5.18 show two possible solutions.

For less symmetric puzzles, the shape of sub-regions can be used to reduce the number of pieces that can fit in that region,

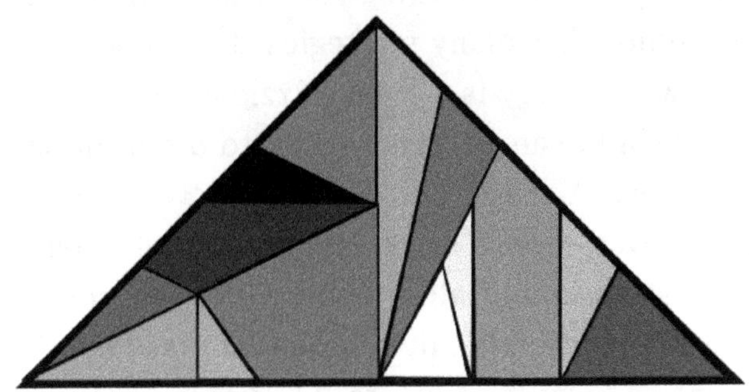

Fig. 5.17   Triangle Ostomachion solution

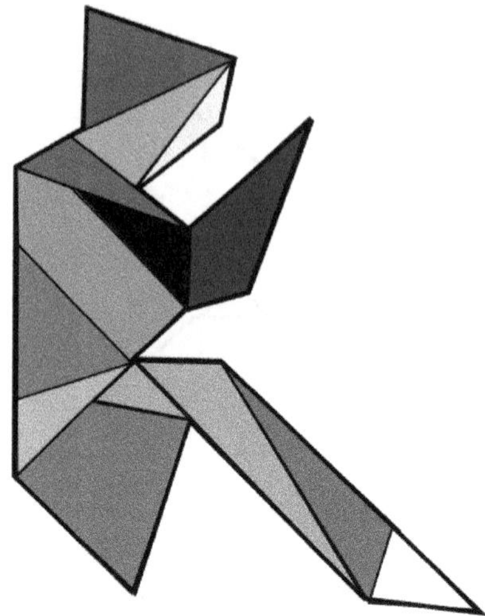

Fig. 5.18   Praying-man Ostomachion solution

making it easier to solve. But even in the square, the corners limit which piece or pieces can go there.

Another strategy frequently used is to begin with the big pieces and work your way down. The number of pieces that can fit into some free space in the puzzle is much bigger if those pieces are smaller.

I heard of a strategy that sounds interesting, but I do not know if or how well it works — and that is to start by filling pieces along the border of the shape before working your way in. Having said all that, there really are not that many strategies that I know of, so coming up with your own strategy is a great puzzle in itself.

The Ostomachion can be generalised to different sets of pieces and more shapes. We are not even sure whether Archimedes' original version was square or rectangular or whether the original puzzle was to re-assemble all 14 parts into the square or create something else. We have already mentioned that Persi Diaconis and his co-workers reduced the set to 11 pieces (Fig. 5.13). Pantazi Houlis, a puzzler designer and the founder and director of the Puzzle

Museum in Megesti, Greece, made a physical version of this 11-piece puzzle. Pantazi is a real puzzle expert and being Greek has studied the palimpsest in depth. He is certain that there is absolutely no connection between the word Ostomachion and the stomach — not even in ancient Greek! If you buy his version of the puzzle, you get with it a nice variation. The puzzle comes with some drawings of squares, each with some red lines that divide it into sub-regions. You must fit the Ostomachion pieces into that square so that no piece crosses a red line. Figure 5.19 shows such a puzzle.

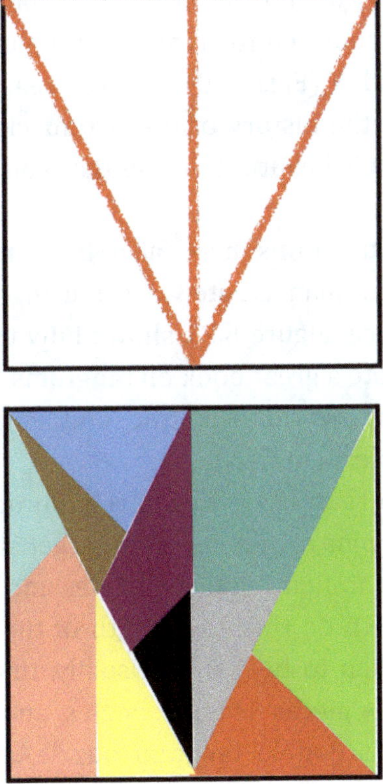

Fig. 5.19   A puzzle that adopts the style of Pantazi Houlis, and its solution. The goal is to fit the 11 pieces in the square without any piece crossing a red line

Kate Jones made her own variation on the Ostomachion. Since her version has the pieces coloured with three different colours, she asked questions like "How many solutions are there with 14 different coloured regions?" A different-coloured region is one where no two shapes of the same colour touch along a shared edge. An example of such a solution is given in her original design (Fig. 5.14). It turns out that there are a total of six solutions. In 2017, Joe Marasco worked on the tricolour Ostomachion and asked a more general question. How many solutions exist for a given number of regions, R, where N same-coloured pieces share an edge or part of an edge? For Kate's original puzzle with 14 regions and no pieces that share an edge, $R = 14$, $N = 0$, and the number of solutions is 6. Kate, Joe, and Alex Streif collaborated to create a catalogue of all the solutions for any given R and N (Fig. 5.20). You can learn much more about their research and the history of the Ostomachion in a monograph that comes with their beautiful puzzle that you can buy on Kadon's website.

Variations on tangrams have also been made. Israeli puzzle designer David Goodman created a set using five of the original tangram pieces twice. Figure 5.21 shows how these can fit into one square. He also wrote a great book on tangrams, *The Tangram Puzzle Book*, published by World Scientific and co-authored by Israeli origamist and puzzler Ilan Garibi.

But there is no reason for you to rely on others. You can make your own set with your own pieces and target shapes.

A nice puzzle is to figure out the angles and areas of each of the pieces of a puzzle. To do this, the length of the edge of the outline of the puzzle is taken to be 1 unit. Usually, that is all you need to know to find out the pieces' lengths, angles, and areas. For example, in David Goodman's double tangram puzzle, assuming the length of the outline square's edge is 1, we can immediately see that the small edges (legs) of the large, blue, isosceles right triangles are $\frac{1}{2}$. How long is the base of these triangles? We need to remember what Pythagoras taught us a few millennia ago.

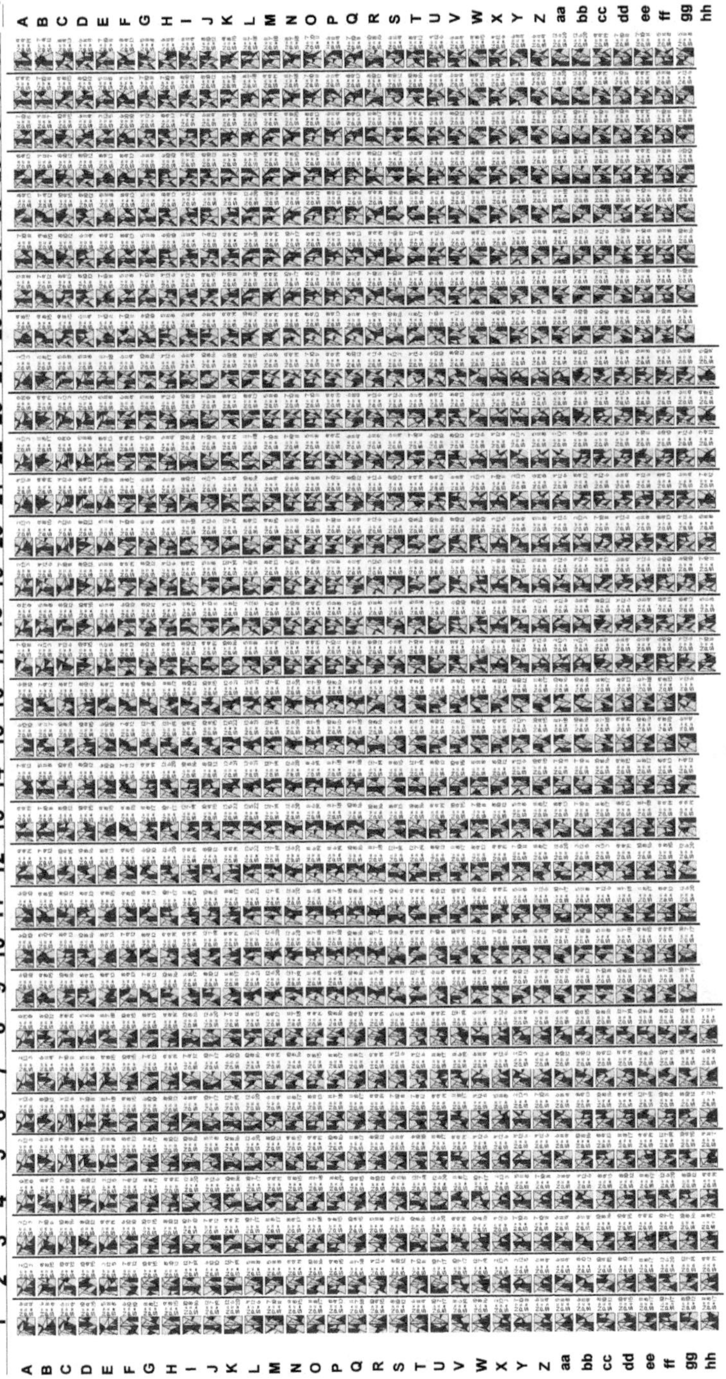

Fig. 5.20   The tricolour Ostomachion catalogue

Fig. 5.21   David Goodman's double tangram

Pythagoras was arguably the world's first philosopher. Well, at least according to the Roman philosopher Cicero, which is what *he* thought of himself. As with most, if not all, ancient Greek philosophers, we know about Pythagoras from the writings of much later, mostly Roman scholars, and so much of what we know, including even what he looked like (Fig. 5.22), is still speculative, given the highly subjective, story-like style of Roman biographers.

He was born in 570 B.C. on the Greek island of Samos, just off the coast of western Turkey, and he moved later on to Croton, now in southern Italy. It was there that he founded a movement — some would call a cult — known as the Pythagoreans or the Pythagorean School. The Pythagoreans believed that reality was mathematical in nature and that philosophy could be used for spiritual purification. They also believed in the soul's immortality, which undergoes a cycle of re-births. They lived as a cult, secluding themselves from society and obeying strict practices, such as not eating meat or beans, maintaining sexual purity, and wearing white clothes. They

Fig. 5.22   This is not what Pythagoras looked like. It is a copy of a
Greek sculpture made centuries after his death, which is on display
at the Musei Capitolini, Rome

had an internal hierarchy, with Pythagoras at the top of their social
structure, with a semi-God status. The pentagram — a five-pointed
star — was their secret sign, used to distinguish themselves and
recognise other members. They believed it was a symbol of power
and immunity. Indeed, they believed in *numbers* as being the essence
of the Universe. "All is numbers," they taught. The number "1"
symbolised unity and the Divine, "2" symbolised duality — thought
and matter, "3" symbolised harmony, and so on. They spent a lot of
time studying astronomy, philosophy, and mathematics, particularly
number theory, geometry, and the philosophy of math. But they also
studied music in great detail, noticing the connection between the
whole number ratios of strings and the harmonies produced when
they are plucked. Being a secluded group with unusual customs, the
Pythagoreans were continuously mocked and persecuted by the
authorities and by (the intolerant) society at large. Pythagoras died
around the year 495 B.C., but the Pythagorean School carried on

despite their persecution for at least a couple of centuries, and their influence carried on well into the first millennium A.D. Nowadays, Pythagoras is widely known for his important theorem on right-angle triangles, which, incidentally, was not found by him, but rather the ancient Babylonians or Egyptians, though he might have been the first to prove the theorem. His theorem, which will help us find the lengths of the double tangram pieces, states that in any right-angle triangle, the sum of the squares of the two legs — the two shorter sides of the triangle — is equal to the square of its hypotenuse — the larger side. This theorem has been proved in so many ways that books have been compiled with the proofs as their sole content! Here is one quick proof. Four right triangles with short legs of length a and b and hypotenuse of length c are placed to form a square with a side length of c surrounding a square with side length $b - a$ (Fig. 5.23).

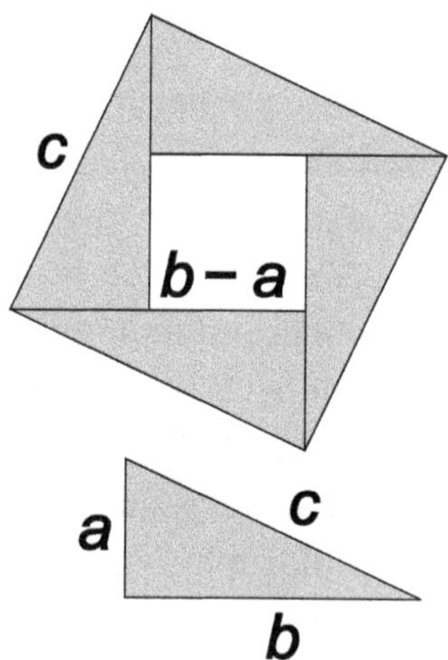

Fig. 5.23   A proof of the Pythagorean theorem

The area of the square of side length c is the product of its side lengths: $c \times c = c^2$, but it is also equal to the sum of the areas of the four triangles and the inner square of side length $b - a$. The area of the inner square is: $(b-a)^2 = b^2 - 2 \times a \times b + a^2$, and the area of a triangle is half the product of its base and its height, so each triangle has an area: $\frac{1}{2} \times a \times b$. Putting this all together, we get: $c^2 = b^2 - 2 \times a \times b + a^2 + 4 \times \frac{1}{2} \times a \times b = a^2 + b^2$, and we are done!

Returning to the double tangram, we saw that the lengths of the legs of large, blue triangles are $\frac{1}{2}$. Using the Pythagorean theorem, we get the length of the base (hypotenuse) of the triangle:

$$\sqrt{\left(\frac{1}{2}\right)^2 + \left(\frac{1}{2}\right)^2} = \sqrt{\frac{1}{2}}.$$

The angle at the corner of this triangle is 90°, and since the angles of a (planar) triangle always sum to 180°, the base angles are 45° each. Work your way down through the green squares (whose edge length is $\sqrt{\frac{1}{8}}$) to get the remaining angles and edges. The areas can be calculated in many ways, but just from close inspection, you can probably see that the area of the large, blue triangle is one-eighth of that of the whole square, i.e. $\frac{1}{8}$. This can be verified using the equation for the area of a triangle — $area = \frac{1}{2} \times height \times base\ length$. For the large, blue triangle, the base length is $\frac{1}{2}$ and the height is $\frac{1}{2}$, so the area is equal to $\frac{1}{2} \times \frac{1}{2} \times \frac{1}{2} = \frac{1}{8}$.

Figure 5.24 shows all the lengths and angles of the double tangram. The areas are:

- *Large blue triangle, green square, orange parallelogram* — $\frac{1}{8}$

- *Small white and purple triangles* — $\frac{1}{16}$

The shapes in the classic tangram have the same dimensions as Goodman's shapes, except that in addition, there are the large right, isosceles triangles, each with a leg length of $\frac{1}{2}$ unit and a base length of 1 unit. Calculating the angles of the Ostomachion shapes is more difficult and is left for the reader as an exercise.

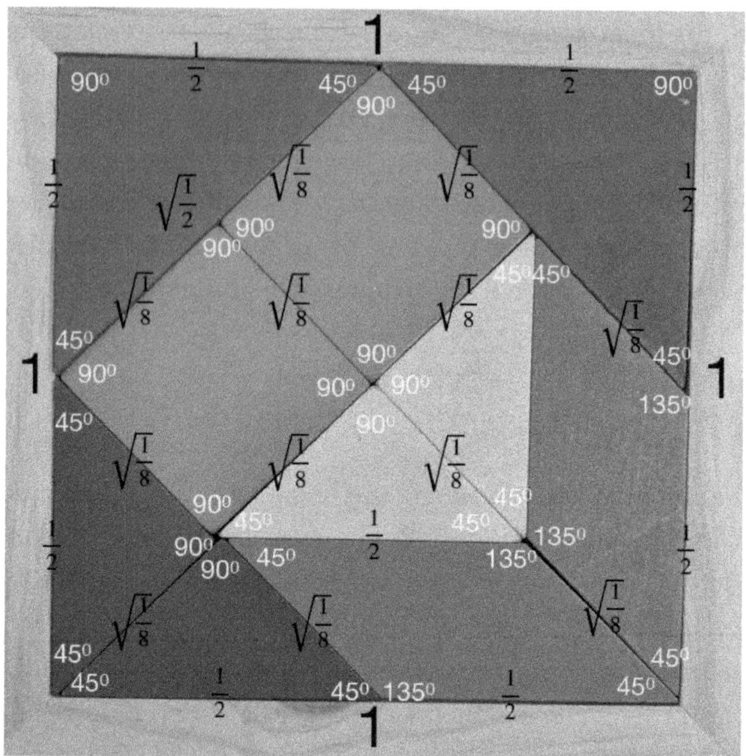

Fig. 5.24   Lengths and angles of the pieces in the double tangram

The most recent generalisation of tangram and Ostomachion puzzles was recently presented at a conference I attended in Atlanta. It was invented by Chicago-based architectural designer and inventor Gilroy Song during COVID-19. Song invented Pickagram, a three-dimensional version of the tangram. You can see the relationship between the seven Pickagram solids and their tangram counterparts in Fig. 5.25.

The solids are magnetic, coloured, multi-faced polyhedra that snap together to make three-dimensional shapes — ducks, ships or whatever (Fig. 5.26).

Each piece has such an unusual shape that just trying to name it is a feat! Song very cleverly used the Golden ratio to build them, a

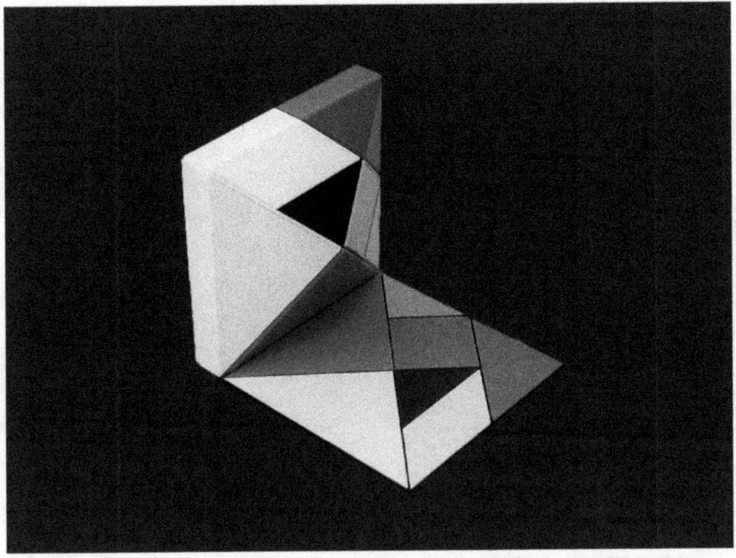

Fig. 5.25 Gilroy Song's Pickagram solids shown as related to their tangram counterparts

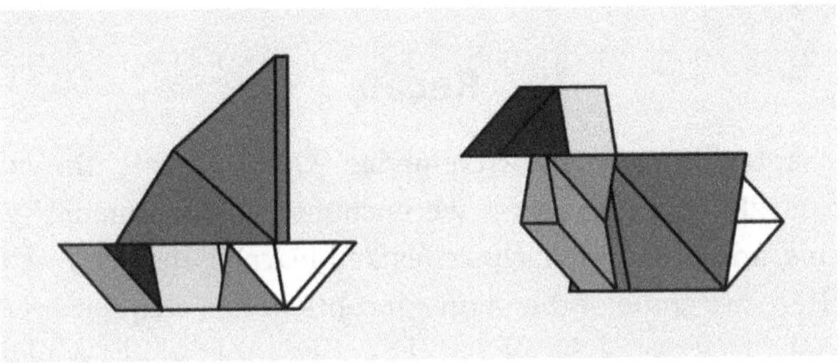

Fig. 5.26 A Pickagram duck and ship

glimpse of which you can see in Fig. 5.27. There is so much more to say about Pickagrams, and being new, there is even more math to be uncovered, but we will have to leave that for another time.

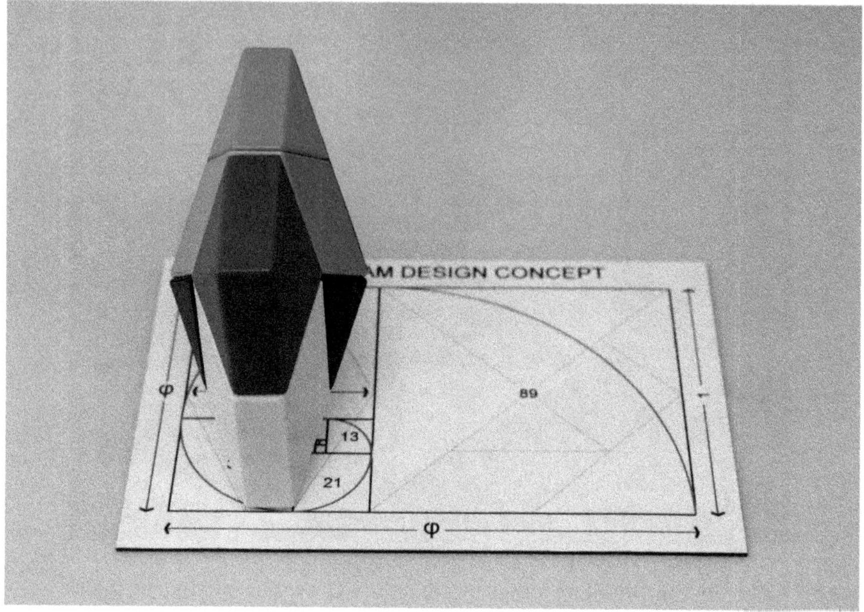

Fig. 5.27   The relationship between Pickagram and the Golden ratio

Recap

This chapter introduced Archimedes' Ostomachion, the world's oldest puzzle. Along the way, we encountered Archimedes' genius, tangrams, and other similar puzzles and learned about how to solve them. Here are some of the main concepts in this chapter:

- *Ostomachion* — an ancient puzzle invented by Archimedes consisting of 14 shapes that can be arranged in 536 different ways to form a square or other geometrical shapes, patterns or pictures. Also known as *Stomachion*.

- *Reduced Ostomachion* — a reduced version of the Ostomachion with 11 shapes.

- *Tangram* — a puzzle similar to the Ostomachion consisting of 7 shapes that can be arranged to form geometrical shapes, patterns, and pictures. There is only one solution where the tangram pieces form a square.

- *Archimedes' screw* — a large screw-shaped contraption used to lift water from a lower to a higher level through a spiral trough that, when turned, lifts the water up and out of an opening near the top.

- *Archimedes' calculation of pi:* $\frac{223}{71} < \pi < \frac{22}{7}$ — the upper and lower limits of pi, the ratio of the circumference of a circle to its diameter. Archimedes calculated these limits by inscribing and circumscribing a circle with regular 96-gons and calculating the ratio of their circumferences to the circle's diameter.

- *Archimedes Palimpsest* — a collection of works from various authors from Ancient Greece, including works by Archimedes.

- *Lucky Puzzle* — a 7-piece tangram-like puzzle made by the Japanese Hanamaya Puzzle Company.

- *Combinatorics* — A branch of mathematics that deals with the number of ways to count things.

- *The Pythagorean theorem* — A theorem that states that the sum of the squares of the two legs of a right triangle is equal to the square of its hypotenuse.

- *Pickagram* — a 3-dimensional version of the tangram.

- *Pick's law (see challenge section)* — the area of a polygon on an equally spaced dotted grid is equal to the number of its interior dots plus half the number of dots on its boundary minus 1.

Challenge Yourself!

(1) Figure 5.28 shows the pieces of the Ostomachion, beautifully crafted by Eli and Galit Gaya.

Fig. 5.28   The 14 Ostomachion pieces

Print them, cut them out, and rearrange them to form (Fig. 5.29):

(a) A kangaroo

(b) A parallelogram

(c) An equilateral triangle

Fig. 5.29   Outlines of a kangaroo, a parallelogram, and an equilateral triangle

(2) Solve the Ostomachion puzzle in Fig. 5.30. You can cut out and use the accompanying pieces. Pieces are not allowed to cross red lines!

Fig. 5.30   Ostomachion puzzle and pieces

(3) Find all the angles and side lengths of the 11-piece Ostomachion, given that the side length of the square is 1 unit, and using the angle given in Fig. 5.31. You may use the green-dashed outline of a rectangle to help.

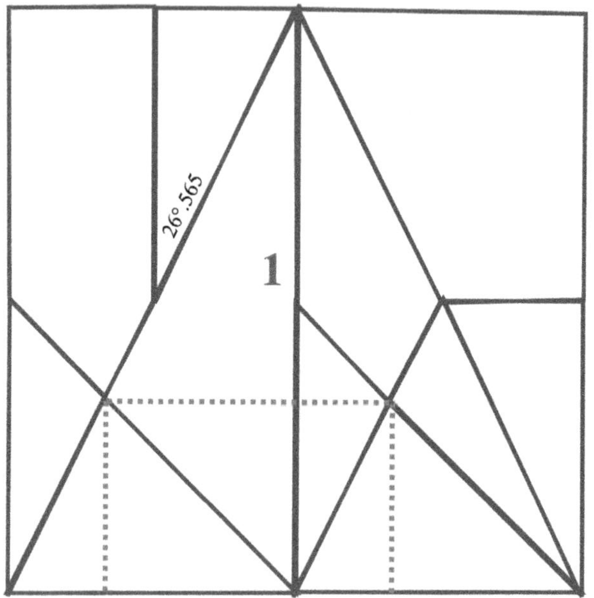

Fig. 5.31   11-piece Ostomachion

(4) Figure 5.32 shows one configuration of Kate Jones' tricolour Stomachion superimposed on a 12×12 grid. The area of one of the pieces is given as 6 units. Find the areas of all the other pieces. Hint: The grid is given so that you can use Pick's law. Pick's law to me is like magic — but it works — and has been proven to be true! It states that the area of a polygon on an equally spaced dotted grid is equal to the number of its interior dots plus half the number of dots on its boundary minus 1. In Fig. 5.32, the area of the given triangle is 6 because it has 3 interior dots and 8 boundary dots, and $3 + 8 \div 2 - 1 = 6$.

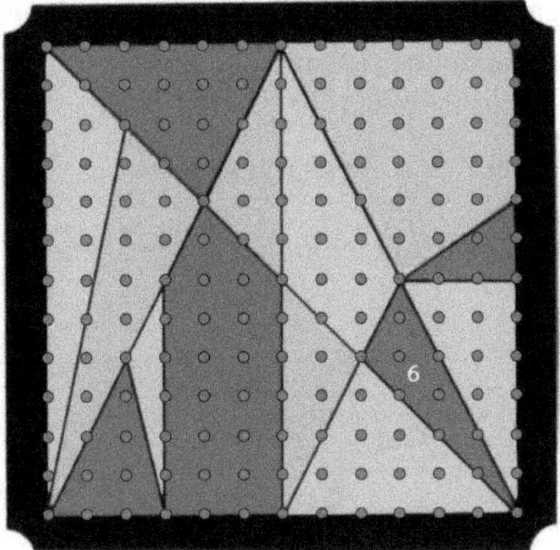

Fig. 5.32　Kate Jones' tricolour Stomachion area puzzle

(5) Figure 5.33 shows a complete 7-piece tangram set. Use six of these pieces to make the largest possible number of squares. Each piece can be used only once (Credit: *The Tangram Puzzle Book*).

Fig. 5.33　A complete tangram set

(6) Use the pieces you used to solve the previous problem to form a square within a square. (Credit: *The Tangram Puzzle Book*).

Solutions

(1) Solutions for the Ostomachion shape puzzles.

Fig. 5.34 Ostomachion solution for a kangaroo

Fig. 5.35 Ostomachion solution for a parallelogram

Fig. 5.36 Ostomachion solution for an equilateral triangle

(2) Solution of the Ostomachion puzzle with red lines.

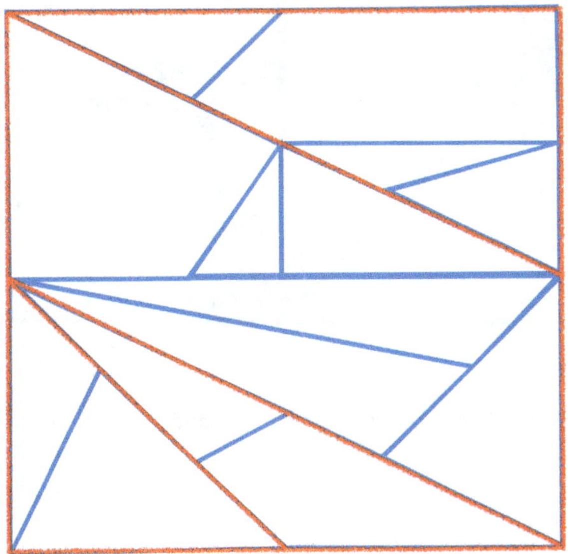

Fig. 5.37 Ostomachion solution for a puzzle with red lines

(3) Solution of Ostomachion angles and side lengths.

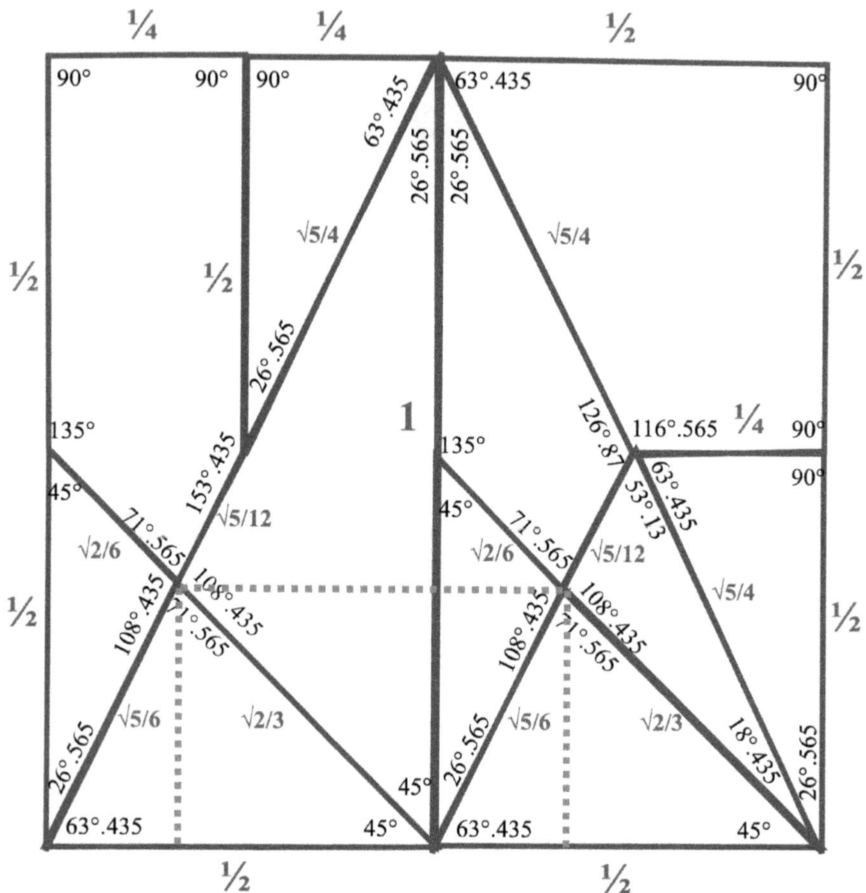

Fig. 5.38   Ostomachion angles and side lengths

(4) Solution of the Ostomachion shape areas puzzle.

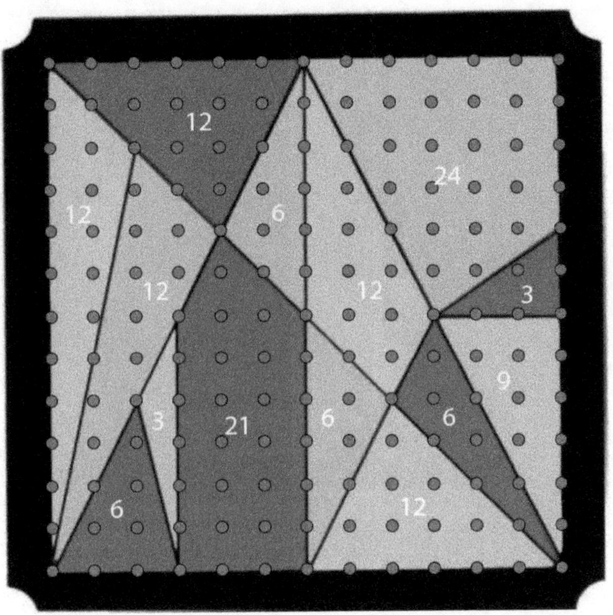

Fig. 5.39   Kate Jones' tricolour Stomachion area puzzle

(5) Solution of the Tangram squares puzzle.

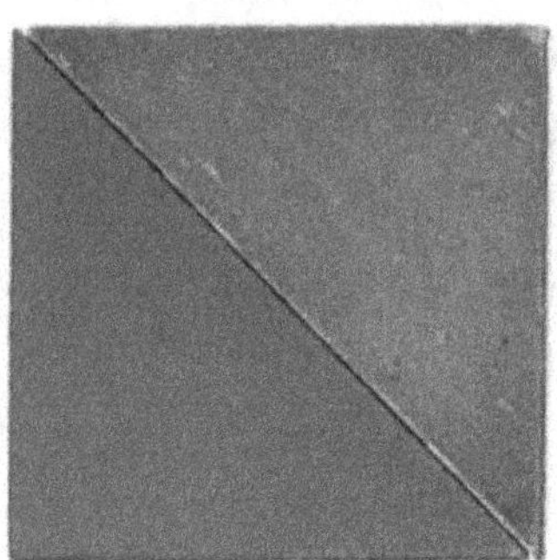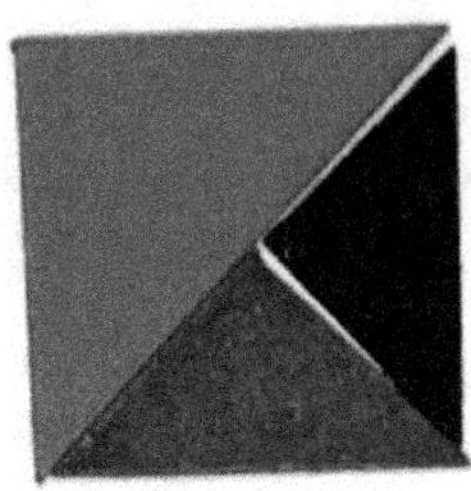

Fig. 5.40   Solution of the tangram squares puzzle

(6) Solution of the square within a square puzzle.

Fig. 5.41   Solution of the square within a square puzzle

Bibliography and Further Reading

Acerbi, F. and Wilson, N. (2004). Towards a reconstruction of Archimedes' Stomachion. *Sciamvs*, **5**: 67–99.

Ausonius, D. M. and Evelyn-White, H. G. (1919). Ausonius, with an English Translation. Harvard University Press. https://archive. org/stream/deciausonius01ausouoft/deciausonius01ausouoft_ djvu.txt. The text is on the public domain.

Ball, W. W. R. and Coxeter, H. S. M. (1987). *Mathematical Recreations and Essays.* Dover Publications.

Chung, F. and Graham, R. A tour of Archimedes' Stomachion. https:// mathweb.ucsd.edu/~fan/stomach/tour/stomach.html

Goodman, D. and Garibi, I. (2018). *The Tangram Puzzle Book.* World Scientific.

Marasco, J., Jones, K., and Streif, A. (2017). *The Tricolor Stomachion.* Independent Publisher.

Mayer, B. (2012). Reproduction of one of *The Parallel Lives by Plutarch*, Vol. V of the Loeb Classical Library edition, 1917. *Greek and Roman Authors on LacusCurtius.* http://penelope.uchicago. edu/Thayer/E/Roman/Texts/Plutarch/Lives/Marcellus*.html. The text is on the public domain.

O'Connor, J. J. and Robertson, E. F. (1999). Pythagoras of Samos. *MacTutor Index.* School of Mathematics and Statistics, University of St Andrews, Scotland. Retrieved December 1, 2023, from https://mathshistory.st-andrews.ac.uk/Biographies/ Pythagoras/.

Pitici, M. (2008). Archimedes' Stomachion. *Geometric Dissections.* https://pi.math.cornell.edu/~mec/GeometricDissections/1.2 Archimedes Stomachion.html/

Sarcone, G. A. and Waeber, M.-J. (2024). Ostomachion, original texts and fragments, *Archimedes' Laboratory.* https://www. archimedes-lab.org/latin_ostomachion.html

Wikipedia contributors. (2023, November 13). Ostomachion. *Wikipedia, The Free Encyclopedia.* Retrieved June 6, 2024 from https://en.wikipedia.org/w/index.php?title=Ostomachion&old id=1184891756.

Wikipedia contributors. (2024, April 25). Archimedes. *Wikipedia, The Free Encyclopedia.* Retrieved June 6, 2024, from https:// en.wikipedia.org/w/index.php?title=Archimedes&oldid= 1220763472.

Wikipedia contributors. (2024, May 27). Tangram. *Wikipedia, The Free Encyclopedia.* Retrieved June 6, 2024, from https://en. wikipedia.org/w/index.php?title=Tangram&oldid=1225985471.

Chapter 6

Diophantine Equations and Mathematical Art

Here is one of my favourite classic puzzles that have been posed in many different ways for as long as antiquity. It can be solved using many methods, one of which is ingenious.

The Puzzle

A farmer has a coop of 30 chickens, some of which have two legs, and the others hop on one leg (don't even ask...). The total number of chicken legs is 47. How many chickens are there of each kind?

Where to Start?

This puzzle *can* be solved using algebra — you know — x for the number of one-legged chickens, y for the number of two-legged chickens, and a system of two equations for the constraints:

$$\begin{cases} x + y = 30 \\ 2x + y = 49 \end{cases}.$$
(6.1)

But! There is a much more clever way to solve the problem using logic and no algebra at all!

Solving the Puzzle

Let's think about the puzzle for a second. Suppose all the chickens had two legs. Then, 60 legs would be counted in total. But only 47 were counted, which is 13 less than 60. So, 13 chickens must have one leg; the rest: $30 - 13 = 17$ are two-legged. We can check that, indeed, this is the correct answer:

$$\begin{cases} 17 + 13 = 30 \\ 2 \times 17 + 13 = 47 \end{cases}. \tag{6.2}$$

Suppose there are two and three-legged chickens instead of one and two-legged chickens, and the rest of the problem remains the same — a total of 30 chickens and 47 legs. Using the same logic, we can assume that if all the chickens were three-legged, the total number of legs would be 90. Since there are 47 legs, this must mean that $(90 - 47) \div 3 = 43 \div 3 = 14\frac{1}{3}$ chickens have 3 legs, and $30 - 14\frac{1}{3} = 15\frac{2}{3}$ chickens have two legs. Oops! We can't cut chickens into thirds. The problem has an extra constraint that we missed: the number of each kind of chicken — two and three legs — must be a whole number. This is the definitive feature of Diophantine equations. So there is no solution to the three-legged/two-legged problem.

The History of the Puzzle and its Inventor

Equations whose unknown variables are whole numbers are known as Diophantine equations, named after Diophantus of Alexandria, a Greek mathematician known as "the father of algebra".

Living for 84 years in the third century A.D., he published a series of 13 books known as *Arithmetica*, comprising 130 algebraic problems. The series was so popular that mathematicians studied it throughout the millennia. It even played an important part in one of the most famous Diophantine equations ever to be solved, Fermat's famous last theorem. The theorem originated as a note that the

French mathematician Pierre de Fermat wrote in the margin of book number 2, and it was only proved in 1995 by British mathematician Andrew Wiles. It states that, although there are infinitely many solutions for the Diophantine equations: $a + b = c$ and $a^2 + b^2 = c^2$, where a, b, and c are whole numbers, there is no solution for powers larger than 2. This means that there is no solution for $a^3 + b^3 = c^3$ or $a^4 + b^4 = c^4$, or generally for $a^n + b^n = c^n$, where n is any whole number larger than 2. I couldn't find any reliable images of Diophantus, but Fig. 6.1 shows page 61 of a 1670 copy of *Arithmetica*, upon which a printed version of Fermat's famous margin note appears. Only the first 6 books of *Arithmetica* still survive, though there have been claims that parts of the seventh book exist within an Arabic manuscript written by someone else. Diophantus also wrote a few other books. Interestingly, his work became known to European mathematicians only during the 16th century when his books were translated from Greek into Latin.

We don't know much about Diophantus' life, and it is even unclear when exactly he was born. We do, however, think that we know how old he was when he died. Legend has it that Diophantus' tombstone was inscribed with the following Diophantine problem: "The first one-sixth of his life was his childhood. After another one-seventh of his life, he got married. Then, after another one-twelfth part of his life, he grew a beard. Five years after that, he became a father. His son lived half as long as he did. Sadly, four years after his son passed away, he also left this world". There's no evidence that this epitaph really appears on his grave; it does appear though in a Greek anthology from the fifth century A.D. If you do the math, using x as the unknown variable, Diophantus' alleged age when he died, you get:

$$\frac{1}{6}x + \frac{1}{7}x + \frac{1}{12}x + 5 + \frac{1}{2}x + 4 = x \qquad (6.3)$$

$$x = \frac{14 + 7 + 12 + 42}{84}x + 9 \qquad (6.4)$$

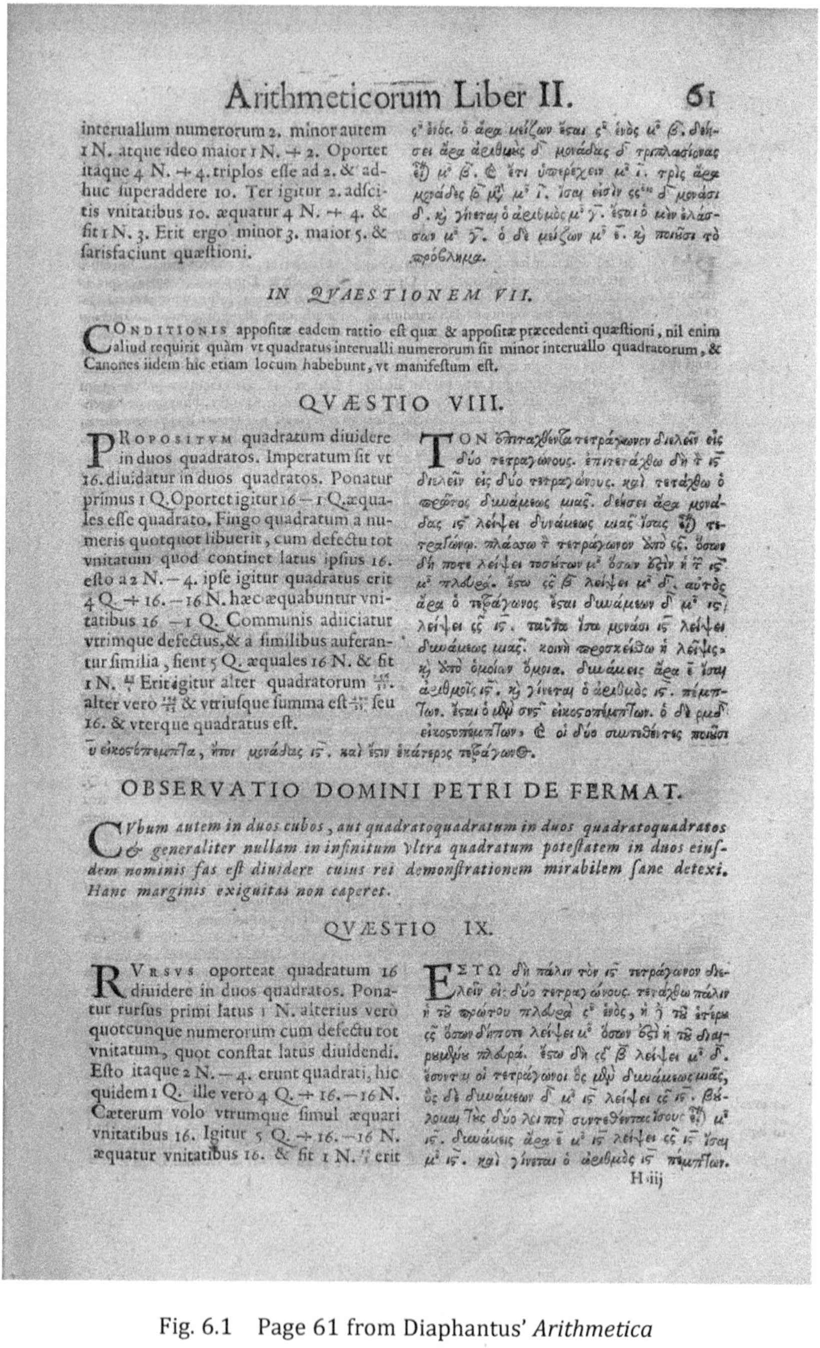

Arithmeticorum Liber II.   61

interuallum numerorum 2. minor autem 1 N. atque ideo maior 1 N. + 2. Oportet itaque 4 N. + 4. triplos esse ad 2. & adhuc superaddere 10. Ter igitur 2. adscitis vnitatibus 10. æquatur 4 N. + 4. & fit 1 N. 3. Erit ergo minor 3. maior 5. & satisfaciunt quæstioni.

IN QVAESTIONEM VII.

CONDITIONIS appositæ eadem ratio est quæ & appositæ præcedenti quæstioni, nil enim aliud requirit quàm vt quadratus interualli numerorum sit minor interuallo quadratorum, & Canones iidem hic etiam locum habebunt, vt manifestum est.

QVÆSTIO VIII.

PROPOSITVM quadratum diuidere in duos quadratos. Imperatum sit vt 16. diuidatur in duos quadratos. Ponatur primus 1 Q. Oportet igitur 16 — 1 Q. æquales esse quadrato. Fingo quadratum a numeris quotquot libuerit, cum defectu tot vnitatum quod continet latus ipsius 16. esto a 2 N. — 4. ipse igitur quadratus erit 4 Q. + 16. — 16 N. hæc æquabuntur vnitatibus 16 — 1 Q. Communis adiiciatur vtrimque defectus, & a similibus auferantur similia, fient 5 Q. æquales 16 N. & fit 1 N. ⅘ Erit igitur alter quadratorum 144/25. alter vero 256/25 & vtriusque summa est 400/25 seu 16. & vterque quadratus est.

OBSERVATIO DOMINI PETRI DE FERMAT.

CVbum autem in duos cubos, aut quadratoquadratum in duos quadratoquadratos & generaliter nullam in infinitum vltra quadratum potestatem in duos eiusdem nominis fas est diuidere cuius rei demonstrationem mirabilem sane detexi. Hanc marginis exiguitas non caperet.

QVÆSTIO IX.

RVrsvs oporteat quadratum 16 diuidere in duos quadratos. Ponatur rursus primi latus 1 N. alterius vero quotcunque numerorum cum defectu tot vnitatum, quot constat latus diuidendi. Esto itaque 2 N. — 4. erunt quadrati, hic quidem 1 Q. ille vero 4 Q. + 16. — 16 N. Cæterum volo vtrumque simul æquari vnitatibus 16. Igitur 5 Q. + 16. — 16 N. æquatur vnitatibus 16. & fit 1 N. ⅘ erit

H iiij

Fig. 6.1   Page 61 from Diaphantus' *Arithmetica*

$$84x = 75x + 756 \tag{6.5}$$

$$9x = 756 \tag{6.6}$$

$$x = 84 \tag{6.7}$$

Diophantine problems were well-known long before *Arithmetica*. The Rhind Papyrus (Fig. 6.2), a famous Ancient Egyptian text containing over eighty math problems from various mathematical topics, included quite a few. Named after Henry Rhind, the Scottish archaeologist who found the papyrus in Luxor, Egypt, in 1858, it was written by a scribe called "Ahmes", and used most probably to train administrative officials in the math they needed for their everyday work.

Fig. 6.2 A fragment of the Rhind Papyrus from the British Museum, London, England

Problem 67, for example, asks the following (I rephrased the problem a bit):

A shepherd had some sheep, two-thirds of one-third of which he had to offer as a tribute to his lord. The shepherd gave 70 sheep. How many sheep did the shepherd originally have?

Since, obviously, the number of sheep has to be a whole number, we have to find the whole number, of which two-thirds of one-third is equal to 70. "Two-thirds of one-third" is two-ninths: $\frac{2}{3} \times \frac{1}{3} = \frac{2}{9}$. Using algebra, we immediately get the answer:

$$\frac{2}{9}x = 70 \tag{6.8}$$

$$x = 70 \times \frac{9}{2} \tag{6.9}$$

$$x = 315 \tag{6.10}$$

The answer is 315 sheep, a third of which is 105 sheep and two-thirds of that is 70, as required. But Ahmes and the Ancient Egyptians didn't know any algebra — it hadn't been invented yet! So, how did they solve the problem? Most probably, they used a method called "false assumption" or, sometimes, "false position". First, you make an initial "wild" guess; for example, guess that the answer to problem 67 is 630.

$$\frac{2}{9} \times 630 = 140 \tag{6.11}$$

This is off by a factor of 2, because we are told in the problem that the shephard gave 70 sheep — not 140:

$$140 \div 70 = 2 \tag{6.12}$$

All you need to do now is to divide your wild guess by this factor to get the correct answer:

$$630 \div 2 = 315 \tag{6.13}$$

This method works in all cases, regardless of whether your guess was too high or too low. If, for example, you guessed 105, then you would get the equation by following these steps:

$$\frac{2}{9} \times 105 = 23\frac{1}{3} = \frac{70}{3} \tag{6.14}$$

$$\frac{70}{3} \div 70 = \frac{1}{3} \tag{6.15}$$

$$105 \div \frac{1}{3} = 315 \tag{6.16}$$

Ancient Egyptian — indeed, math from all ancient cultures — is very interesting. We know how the Egyptians did math from several papyruses that survived nearly four millennia! The Rhind Papyrus is estimated to have been written around 1650 B.C. The Moscow Papyrus, the Berlin Papyrus 6619, and the Kahun Papyrus date even earlier, and there are others as well. Although the Ancient Egyptians didn't know the decimal system and the concept of zero was unheard of (to be discovered roughly 2,000 years later), their number system was nevertheless based on powers of ten. They used one of three notations: hieroglyphics, hieratics and demotics. Figure 6.3 shows a table of the units, tens, hundreds, and thousands notations for the digits 1–9. The table was compiled by Kurt Sethe, a German Egyptologist, who published a comprehensive history of mathematical notations in 1928.

The Egyptians didn't do arithmetic the way we do now. They added and subtracted numbers by counting the symbols, replacing them with the next level ones whenever they reached ten. The number 4, for example, would be drawn as four strokes in hieroglyphics. The number 17 would be drawn as a cattle hobble (looks like an upside-down horseshoe) that notates the number 10 and seven strokes. Adding 17 and 4 (they drew left-facing and right-facing footsteps in the Rhind Papyrus to denote addition and

	Einer			*Zehner*			*Hunderte*			*Tausende*			
1													
2													
3													
4													
5													
6													
7													
8													
9													
	hierogl.	hierat.	demot.	hierogl.	hierat.	demot.	hierogl.	hierat.	demot.	alter hierogl.	jünger hierogl.	hierat.	demot.

Fig. 6.3   Kurt Sethe's compilation of Ancient Egyptian numerals

subtraction), you get 21, so they would perhaps originally draw 11 strokes and one hobble, and then replace ten of the strokes with another hobble to get two hobbles and one stroke.

Multiplication was done as multiple additions, though they did this using an ingenious algorithm. They first wrote down the multiplicand, and next to it, they wrote, in a separate column that we'll call for short the "multiplier column", the number 1. They then repeatedly doubled both the multiplicand *and* the number in the multiplier column, writing down the results in separate rows until they got a number larger than half the multiplier in the multiplier column. They then figured out which combination of numbers in the rows of the multiplier column adds up to the multiplier and added the numbers in the respective rows of the multiplicand column to get the final answer.

Table 6.1 is an example of how they calculated 17×13. First, they wrote down 17, the multiplicand; next to it, they wrote the number 1. Then, they doubled both rows repeatedly, writing down the results each time until they got a number larger than $6\frac{1}{2}$ in the multiplier column — in our case, 8. Noticing that $1 + 4 + 8 = 13$, they added the numbers in the corresponding rows in the multiplicand column, the shaded rows in the table, to get the correct answer: $17 + 68 + 136 = 221 = 17 \times 13$.

Table 6.1　Multiplying 17×13 using the Egyptian multiplication method

	Multiplicand Column	Multiplier Column
	17	1
	34	2
	68	4
	136	8
Total (Shaded Rows)	**221**	**13**

Division was done in a similar manner. To divide 221 by 17, you would double 17 repeatedly until you got the largest number less than 221, which would be 136. You would then find those rows whose numbers add up to 221 and locate the corresponding numbers in the column next to it. The sum of these numbers is the correct answer. With division, of course, the answer might not be a whole number, in which case fractions need to be added and calculated.

Egyptian fractions are an interesting topic in their own right. The Egyptians knew only three kinds of fractions: the fraction $\frac{2}{3}$, the fraction $\frac{3}{4}$, and fractions where the numerator is 1, like $\frac{1}{2}, \frac{1}{3}, \frac{1}{4}, \frac{1}{5},$

and so on. The reason for this is probably because arithmetic was used for practical purposes. When you want to divide up a whole pie for some kids, it's easy to do this by cutting the pie into whole number slices, 2, 3, 4, 5, etc., to get $\frac{1}{2}$ pie each if there are two kids, $\frac{1}{3}$ pie each for three kids, $\frac{1}{4}$ for four, $\frac{1}{5}$ for five, and so on. If, for some reason, you need to cut up the pie into $\frac{1}{3}$ and $\frac{2}{3}$, or $\frac{1}{4}$ and $\frac{3}{4}$, you probably wouldn't have much trouble. But for other fractions, like $\frac{7}{8}$ or $\frac{5}{6}$, things might get a bit tricky. Fractions where the numerator is 1 are known as *unit fractions*.

The Egyptians' fraction calculations are actually far more advanced than you might think. We know from Ahmes's papyrus and others that the Egyptians liked to write fractions as a sum of unit fractions where each fraction is *distinct*. This means that instead of writing $\frac{5}{8}$ as $\frac{1}{8} + \frac{1}{8} + \frac{1}{8} + \frac{1}{8} + \frac{1}{8}$, they would write it as $\frac{1}{2} + \frac{1}{8}$. Finite sums of distinct unit fractions are known as *Egyptian fractions*. Egyptian fraction mathematics has always been a popular topic for mathematicians to study, and they still do so today. There are even some *open problems* — problems that mathematicians have not been able to solve yet — about Egyptian fractions. One of them, known as the Erdős–Straus conjecture, is: can every fraction whose numerator is 4 be written as an Egyptian fraction that is a sum of three unit fractions at the most?

Some Egyptian fractions are easy to guess. Others are much harder. I know a nice and relatively easy method to find Egyptian fractions, attributed to Italian mathematician Leonardo Fibonacci, who wrote about it in his famous math book *Liber Abaci* in 1202. Take the fraction you want to change into an Egyptian fraction and turn it upside down, making the numerator the denominator and vice versa. For example, to calculate the Egyptian fraction $\frac{11}{19}$, first write $\frac{19}{11}$. Then, find the first whole number larger than this fraction. For $\frac{19}{11}$, this is 2. The first unit fraction of the Egyptian fraction we are looking for is one divided by this whole number — in our case, $\frac{1}{2}$. To find the next unit fraction, we find the difference between our original fraction, $\frac{11}{19}$, and the first unit fraction, $\frac{1}{2}$, so, $\frac{11}{19} - \frac{1}{2} = \frac{22}{38} - \frac{19}{38} = \frac{3}{38}$. We now do the same with this fraction: turn it upside down to get $\frac{38}{3}$, find the next largest whole number, 13, and this becomes the denominator of the next fraction in the series, $\frac{1}{13}$. We now have $\frac{11}{19} = \frac{1}{2} + \frac{1}{13} + \ldots$ We still haven't finished. Now, we subtract both terms from $\frac{11}{19}$ and get $\frac{11}{19} - \frac{1}{2} - \frac{1}{13} = \frac{286 - 247 - 38}{494} = \frac{1}{494}$. Yay! $\frac{1}{494}$ is a unit fraction, so we can safely write $\frac{11}{19} = \frac{1}{2} + \frac{1}{13} + \frac{1}{494}$. Fibonacci's algorithm works for all rational numbers but doesn't necessarily find the optimal way to write a fraction. That means that other algorithms might find shorter expansions. It is, what one might call "the long and sure way round" — that's why it's called a "greedy algorithm". When done on a computer, it can take up a lot of resources. Still, it's straightforward and fairly easy to work with.

We seem to have strayed from whole numbers to fractions, so let's return to the history of Diophantine equations, travelling further up the timeline, but still before Diophantine's lifetime, to Archimedes. In Chapter 2, we briefly mentioned Archimedes' "cattle problem" from the third century B.C. The problem is divided into two parts. The first goes as follows. Many white, black, yellow, and spotted cows and bulls were sacrificed to the Sun God throughout many centuries.

- The number of white cows was the same as $\frac{7}{12}$ the number of black cattle.

- The number of black cows was the same as $\frac{9}{20}$ the number of spotted cattle.

- The number of spotted cows was the same as $\frac{11}{30}$ the number of yellow cattle.

- The number of yellow cows was the same as $\frac{13}{42}$ the number of white cattle.

- The number of white bulls was the same as the number of yellow bulls plus $\frac{5}{6}$ the number of black bulls.

- The number of black bulls was the same as the number of yellow bulls plus $\frac{9}{20}$ the number of spotted bulls.

- The number of spotted bulls was the same as the number of yellow bulls plus $\frac{13}{42}$ the number of white bulls.

Each of the seven statements is an equation with unknown quantities (in mathematical terms, variables): the number of black bulls, white bulls, spotted cows, etc. What's the solution? We can write equations to replace the seven statements, using uppercase for bulls, lowercase for cows, and w, b, y, and s for white, black, yellow, and spotted, respectively:

$$W = \frac{5}{6}B + Y \tag{6.17}$$

$$B = \frac{9}{20}S + Y \tag{6.18}$$

$$S = \frac{13}{42}W + Y \tag{6.19}$$

$$w = \frac{7}{12}(B + b) \tag{6.20}$$

$$b = \frac{9}{20}(S + s) \tag{6.21}$$

$$s = \frac{11}{30}(Y + y) \tag{6.22}$$

$$y = \frac{13}{42}(W + w) \tag{6.23}$$

Wait a minute! There are only seven equations, yet there are *eight* unknowns! This is known as an indeterminate set of equations, meaning we can choose any of the unknown variables and rewrite the seven equations as a function of this unknown. It also means that the problem has an infinite number of solutions. Rewriting these equations so that the number of each type of cattle depends only on the number of yellow bulls can be done using a bit of algebra. We get:

$$W = \frac{10366482}{4149387}Y \tag{6.24}$$

$$B = \frac{7460514}{4149387}Y \tag{6.25}$$

$$S = \frac{7358060}{4149387}Y \tag{6.26}$$

$$w = \frac{7206360}{4149387}Y \tag{6.27}$$

$$b = \frac{4893246}{4149387}Y \tag{6.28}$$

$$s = \frac{3515820}{4149387}Y \tag{6.29}$$

$$y = \frac{5439213}{4149387}Y \tag{6.30}$$

Recalling that these are Diophantine equations, meaning the unknown variables must be whole numbers, even the solution with the minimal amounts of cattle involves very large numbers:

10,366,482 White bulls,
7,460,514 Black bulls,
7,358,060 Spotted bulls,
4,149,387 Yellow bulls,
7,206,360 white cows,
4,893,246 black cows,
3,515,820 spotted cows,
5,439,213 yellow cows.

That's a lot of cattle for the Sun God! The *infinite* set of solutions can be generated by multiplying these numbers by the whole numbers 1, 2, 3, etc., so the second minimal solution would be $10366482 \times 2 = 20732964$ white bulls, $7460514 \times 2 = 14921028$ black bulls, etc.

The second part of Archimedes' cattle problem is even harder to solve. Using the same relationships between the numbers of the different cattle [Eqs. (6.17)–(6.23)], Archimedes now requires that the total number of black and white bulls be a square number and that the total number of yellow and spotted bulls be a triangular number. Square and triangular numbers were discussed in the chapter on number pyramids in *Lewis Carroll's Cats and Rats ... and Other Puzzles with Interesting Tails*, so we'll just give a quick definition now.

Square numbers are integers multiplied by themselves, like $2^2 = 2 \times 2 = 4$, $91^2 = 91 \times 91 = 8281$, $(-6)^2 = (-6) \times (-6) = 36$, etc. Triangular numbers are half the product of two consecutive integers, like $\frac{1}{2} \times (2 \times 3) = 3$, $\frac{1}{2} \times (-10 \times -11) = 55$, $\frac{1}{2} \times (26 \times 27) = 351$, etc. The numbers that satisfy Archimedes' extra conditions are mind-bogglingly large and were only explicitly calculated in 1880 by Carl Ernst August Amthor, the headmaster of the Gymnasium of the Holy Cross in Dresden, Germany. He found that the total number of cattle is humungous — a 206,545-digit number. Mathematicians would *approximate* this number by keeping only the two most significant digits — the first two digits on the left — and then write the number down this way: 7.76×10^{206544}, that is, 77,600,000... carry on writing another 206,537 zeroes. Writing large numbers this way is called *scientific notation*. There are two problems with scientific notation. First, the number is inaccurate; it's only a rounded approximation. The true number begins with 776 but does not continue with a string of zeroes. The shorthand mathematical expression is accurate enough for practical calculations, certainly for a recreational math problem, but in other cases, every digit can be crucial, even lifesaving. A report for the U.S. Subcommittee on Investigations and Oversight, Committee on Science, Space, and Technology disclosed that a fatal Iraqi rocket attack on Dhahran, Saudi Arabia, in 1991, was not intercepted by the Patriot missile defence system because the computer system could not store enough digits of a large number.

The second problem with scientific notation is psychological. Writing a large number using a small number of digits gives the impression of a smaller number than it really is. The 11 digits in 7.76×10^{206544} don't do justice to the true giganticness of the number. Now, I'm a bit finicky about these things and would *really* like to write down the full number. In my previous book, I wrote down a 2,184-digit number using up a page and a half. However, it would take about 150 pages to write the exact number of cattle, so I thought of a nice way to display the full number without taking up much space. Here's how it's done. Rewrite the number in a grid, one digit per cell, and then colour each digit differently. This is exactly what I did for Fig. 6.3. I rewrote the number in a 505×409 rectangular grid and then coloured the 0's light blue, the 1's dark blue, the 2's light green, etc. The picture (Fig. 6.4) is an accurate description of the solution to Archimedes' second cattle problem.

The pictorial depiction of large numbers has come to be one of the many connections between art and math. Martin Krzywinski, a staff scientist at Canada's Michael Smith Genome Sciences Centre, is an expert in combining art and numbers, and his works have appeared in journals, books, and exhibitions all over the world. Figure 6.5 shows his depiction of the first 13,689 digits of pi, generated in a spiral from the centre outwards.

One of the reasons to represent numbers this way is to spot hidden patterns that might give a clue to some interesting math lurking behind the seemingly random digits.

Unfortunately, no such pattern has been found in the digits of pi; however, if you draw the whole numbers in a spiral and mark only

Fig. 6.4   Pictorial solution of Archimedes' second part of the cattle problem

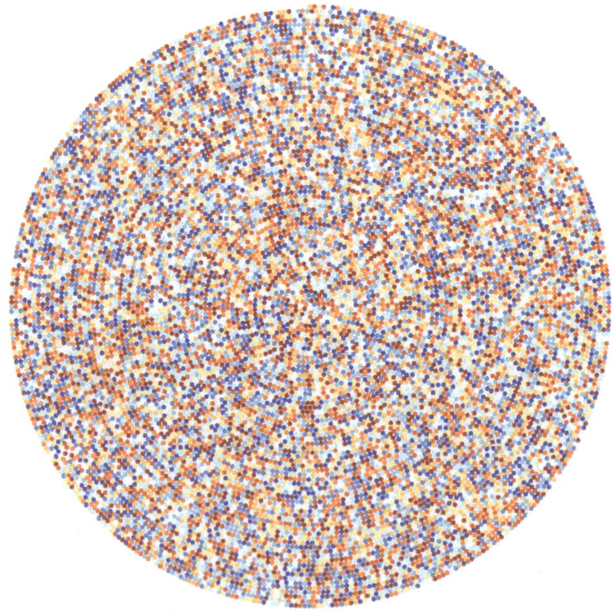

Fig. 6.5   The first 13,689 digits of pi. Image credit: Martin Krzywinski

those that are prime (divisible only by themselves and 1, except 1), then an unexpected pattern of diagonal marks surprisingly emerges. This image, shown in Fig. 6.6, is known as "Ulam's spiral", named after the Polish-American Stanislaw Ulam, who discovered it in 1963. Mathematicians have found explanations for some of the features in Ulam's spiral, but a definitive explanation has yet to be found.

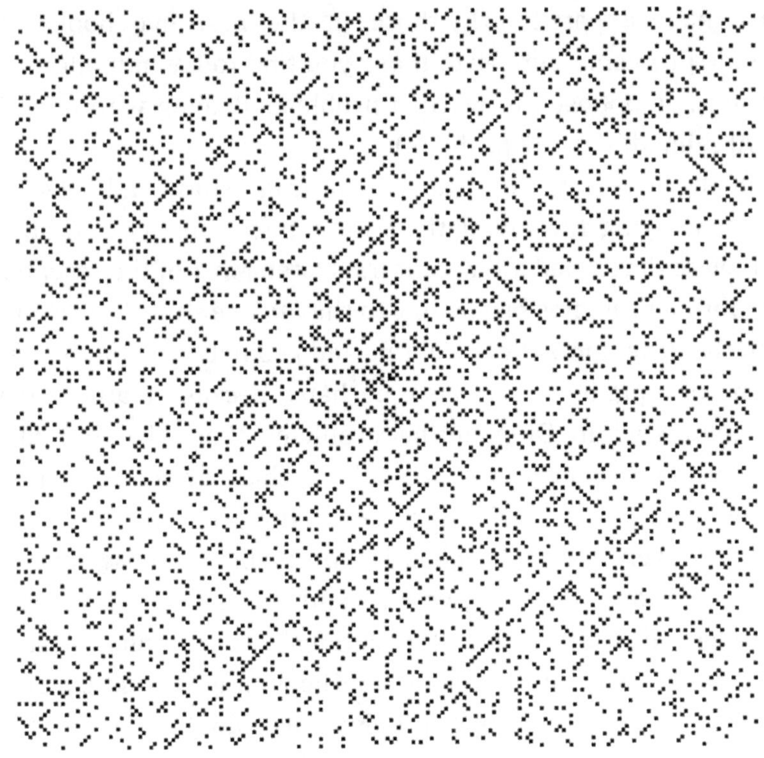

Fig. 6.6　Ulam's spiral

If Diophantine equations had been known and explored long before Diophantus' time, what was so important about his *Arithmetica* book series? First, Diophantine was the first to introduce a special notation known as syncopated notation, which included mathematical symbols. Before this, equations and mathematical operations were described and solved using words and phrases. Even after his time, mathematical symbols were rarely used, except for the very basic arithmetic operations like adding and subtracting. In this aspect, Diophantine was clearly very much ahead of his time.

Second, he was one of the first mathematicians to regard numbers, including fractions, as an abstract concept rather than something concrete that is useful in daily life. Most of the problems in *Arithmetica*, particularly those in the first five books, were algebraic problems, similar to those you find today in modern math textbooks, as opposed to worded problems dealing with quantities (like Archimedes' cattle problem).

Finally, he was the first to *systematically* study and solve these equations, laying the foundations for number theory and algebra, which were developed a few centuries later. Most of the research on Diophantine problems began only in the 20th century. This is mainly due to the development of algebraic and geometric methods to solve systems of equations. It is also, perhaps, because mathematicians were interested in Diophantine problems that, like Archimedes' cattle problem, have fewer equations than unknown variables, making them (arguably) more interesting for mathematicians. Wikipedia even *defines* Diophantine problems as those that "have fewer equations than unknowns and involve finding integers that solve simultaneously all equations". This implies that they would not classify the puzzle at the beginning of this chapter, or the Diophantine gravestone problem, as Diophantine problems. If we want to be finicky, we would probably define a Diophantine equation as a single equation, as opposed to a Diophantine problem that refers to a system of equations, though these definitions are not rigorously obeyed in the literature.

The fact that there are fewer equations than unknowns in Diophantine problems does not necessarily mean they have infinitely many solutions. Finding if and how many solutions exist is part of Diophantine analysis, a field of mathematical research that is still active today.

Generalisations of the Puzzle and Connections to Other Math Areas

Throughout the years, several Diophantine problems have become "famous", and each of them leads to interesting areas of math. We've already mentioned one famous Diophantine equation — Fermat's last theorem. The next is as old as Ahmes. Remember the Pythagorean theorem from Chapter 5? The theorem states that the sum of the squares of a right triangle's two legs equals the square of its hypotenuse (Fig. 6.7). The theorem's algebraic form is well-known:

$$a^2 + b^2 = c^2 \tag{6.31}$$

When we restrict the solutions to whole numbers, the equation becomes a Diophantine equation, and the whole-number solutions to this equation are known as Pythagorean triples. (3, 4, 5) is a Pythagorean triple because $3^2 + 4^2 = 5^2 \rightarrow (9 + 16 = 25)$. Multiplying a Pythagorean triple by any number still satisfies the equation, so each Pythagorean triple leads to an infinite number of triples. So, multiplying the set of numbers (3, 4, 5) by 2 leads to the Pythagorean triple: (6, 8, 10), and indeed, $6^2 + 8^2 = 10^2 \rightarrow$

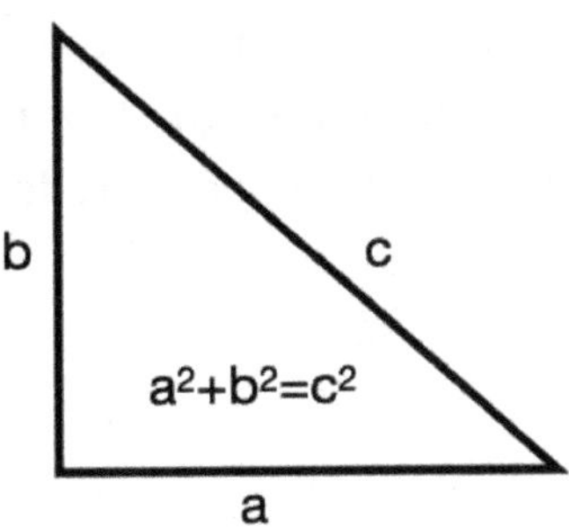

Fig. 6.7   The Pythagorean theorem

$(36 + 64 = 100)$. Multiplying it by three gives another Pythagorean triple (9, 12, 15), and so on, creating a series of Pythagorean triples all related to the initial (3, 4, 5) triple. Mathematicians distinguish between the first Pythagorean triple in a series and its multiples, calling it a *primitive Pythagorean triple.* The first four primitive Pythagorean triples are (3, 4, 5), (5, 12, 13), (8, 15, 17), and (7, 24, 25).

The famous Greek mathematician Euclid gave a simple formula for generating primitive Pythagorean triples. First, you need two positive, *coprime*, whole numbers, with exactly one of them an even number. Coprimes are numbers with no common divisor except for the number 1. 5 and 14 are coprimes because no whole number divides both 5 and 14 without remainder. 9 and 12 are not coprimes because they have a common divisor, 3, that is not 1. Here's Euclid's method for generating a primitive Pythagorean triple. We'll pick the numbers 4 and 7 for our example. Take the larger of the numbers, square it, and subtract the square of the second number. This is the first number of the triple. In our example, $7^2 - 4^2 = 49 - 16 = 33$. The second number is twice the product of the two numbers that we picked. In our example, $2 \times 4 \times 7 = 56$. The third number is, of course, the square root of the sum of the squares of the first and second numbers *of the triple.* So, in our case, it would be the square root of $33^2 + 56^2 = 1089 + 3136 = 4225$. The square root of 4,225 is 65, so the primitive triple is (33, 56, 65). Euclid's formula generates all primitive triples — there are infinitely many.

Figure 6.8 shows a beautiful, pictorial representation of all the Pythagorean triples, where both a and b are less than 100, including negative a and b greater than -100. The picture is a graph with the axes removed, so the center of the picture is the point (0, 0). Each point represents an (a, b) pair of one Pythagorean triple. Once again, we can see the power of mathematical art. In one shot, we can see a lot of detail and some interesting phenomena, and we also get a feeling of the number of triples within the given range.

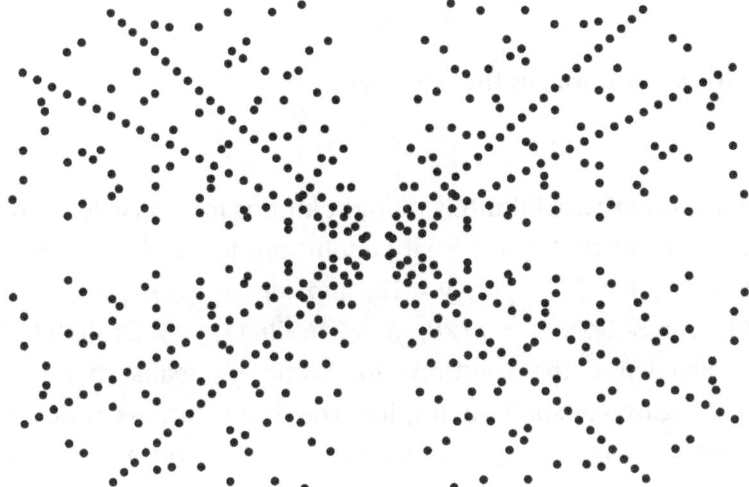

Fig. 6.8   Pictorial representation of the Pythagorean triples

Take the diagonal rays, for example. These exemplify the fact that primitive Pythagorean triples can be multiplied by any whole number to get a new Pythagorean triple. There is a wealth of math on Pythagorean triples — just peek at the length of the Pythagorean triple article on Wikipedia.

Another ancient Diophantine equation, Pell's equation, was studied by Diophantus himself, as well as many other prominent mathematicians — Euclid, Brahmagupta, Brouncker, Fermat, and, arguably, British mathematician John Pell. Pell "lent" his name to the equation but essentially only published the work of others. Pell's equation is really a whole bunch of equations, all of which have the same basic structure:

$$x^2 - ny^2 = 1 \qquad\qquad (6.32)$$

where x and y are the sought unknowns and n is a whole number that is not a square number — so n can be 2 or 3 or 45 or 9,113, but not 1 or 4 or 36 or 4,225. For example, taking $n = 2$, we get the equation:

$$x^2 - 2y^2 = 1 \qquad (6.33)$$

and with $n = 45$, we get the equation:

$$x^2 - 45y^2 = 1 \qquad (6.34)$$

It turns out there are infinitely many solutions for every Pell equation of this form. The first three, positive solutions for Eq. (6.33) are: $x = 1$, $y = 0$; $x = 3$, $y = 2$; $x = 17$, $y = 12$; and for Eq. (6.34) are: $x = 161$, $y = 24$; $x = 51841$, $y = 7728$; $x = 16692641$, $y = 2488392$. Once again, visualising the solutions for some n's leads to important insights. Figure 6.9 shows a graph of the first solutions of Eq. (6.33), including negative (x, y). The solutions all fall on a *hyperbola* — a curve that looks like two horizontal U's back to back. All Pell equations (any n) look like hyperbolas.

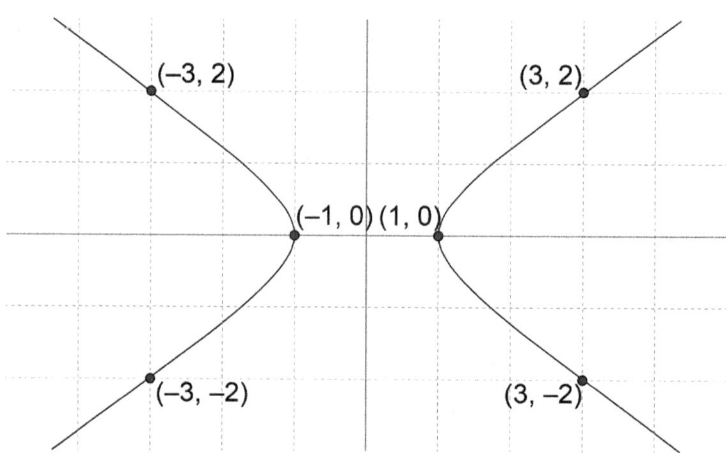

Fig. 6.9   Graph of the first solutions to Pell's equation with $n = 2$

The *abc conjecture* is yet another Diophantine problem that mathematicians have yet to prove, although Japanese mathematician Shinichi Mochizuki claims to have done it — and his proof is still under rigorous and sceptical analysis. Indeed, we use the word *conjecture* to mean a theorem that has not yet been proved. The problem was put forward in 1985 by French mathematician Joseph Oesterlé and British mathematician David Masser, so it is also known as the Oesterlé–Masser conjecture. The problem concerns the very simple Diophantine equation: $a + b = c$, where a, b and c are all positive whole numbers, *and* they are coprime. Recall that this means that no single whole number divides them all. Example: (1, 8, 9) and (3, 5, 8) are coprime triples that satisfy the simple equation $a + b = c$. We'll use these two triples in a minute to demonstrate the conjecture. But before that, we need to define one more thing — the *radical*. The *radical* of a given number is the product of its *distinct prime factors*, meaning the product of all the prime numbers that divide it without remainder. The radical is symbolised by the three letters: *rad*.

Here are two examples of radicals.

- $rad(72) = 6$ because the two prime numbers that divide 72 are 2 and 3, and their product is: $2 \times 3 = 6$.

- $rad(120) = 30$ because the prime numbers that divide 120 are 2, 3, and 5, and their product is: $2 \times 3 \times 5 = 30$.

The abc conjecture concerns the radical of the product of the triple (a, b, c), which for the two examples above is $rad(1 \times 8 \times 9) = rad(72) = 6$, and $rad(3 \times 5 \times 8) = rad(120) = 30$. Notice that c — the number 9 — in the triple (1, 8, 9) is greater than $rad(1 \times 8 \times 9) = 6$, but c — the number 8 — in the triple (3, 5, 8) is smaller than $rad(3 \times 5 \times 8) = 30$. It turns out that this is the more common case, i.e. more often than not, c is smaller than $rad(a \times b \times c)$.

There are infinitely many triples where $c < rad(a \times b \times c)$. There are also infinitely many triples where $c > rad(a \times b \times c)$ — even though they are much rarer — and (1, 8, 9) is one of them. Let's zoom in on this triple for a moment. What happens if we raise $rad(1 \times 8 \times 9)$ to powers greater than 1? Is 9 still greater than these "powers of radicals"? Our intuition tells us that for small numbers, yes, and for large numbers, no. You can use your calculator and experiment if you like. Take a small power like 1.001; we get $rad(1\times8\times9)^{1.001} = 6^{1.001} = 6.011...$ only marginally greater than 6 and definitely smaller than 9. Take a larger power, like 3; $rad(1\times8\times9)^3 = 6^3 = 216$. That's much greater than 9. Let's continue our experiment to try and "nail" the range for which 9 is greater than the power of the radical of $(1\times8\times9)$.

Formally, we ask: when is 9 greater than $rad(1\times8\times9)^{1+k}$, where k stands for a number greater than 0? There is a mathematically subtle reason we use the expression $1+k$ instead of just k, but we won't bother with that now.

- When $k = 0.1$, then $rad(1\times8\times9)^{1.1} = 6^{1.1} = 7.177...$ smaller than 9.

- When $k = 0.2$, then $rad(1\times8\times9)^{1.2} = 6^{1.2} = 8.586...$ smaller than 9.

- When $k = 0.5$, then $rad(1\times8\times9)^{1.5} = 6^{1.5} = 14.697...$ which is now larger than 9.

We could carry on forever. Luckily, mathematicians have found the range for many triples, including ours. It turns out that for (1, 8, 9), $9 > rad(1\times8\times9)^{1+k}$ when $k < 0.226...$ We can say that (1, 8, 9) is one triple for which there are solutions where $c > rad(a\times b\times c)^{1+k}$. Another triple is (1, 48, 49). $c > rad(a\times b\times c)^{1+k}$ is satisfied for this triple when $k < 0.103$.

How many triples like this are there? Just the two we found here? A hundred? A thousand? A Googleplex? Perhaps an infinite number of triples? Mathematicians don't yet know, and to date, they have found over 23 million triples, but they are pretty sure that there is no infinite number of such triples. That is, they *conject* that there is only a *finite* number of triples where c is larger than that number. And that, ladies and gentlemen, is the abc conjecture. It hasn't been proven yet, so maybe one of you, one of these days, will prove it, at least unless Mochizuki's proof is approved — or some other mathematician gets to it before you do!

It's worth mentioning that the abc conjecture, although relatively simple to understand, has huge implications for math. If someone does find the proof, it will solve a whole bunch of problems that currently trouble mathematicians.

The last thing that I'd like to mention about the abc conjecture is that, here, too, creating a pictorial image of those triples that satisfy the conjecture gives us much insight into the problem. Figure 6.10 shows a plot of the upper bound of $1 + k$ for all c's up to 2,000. What do *you* learn from the graph?

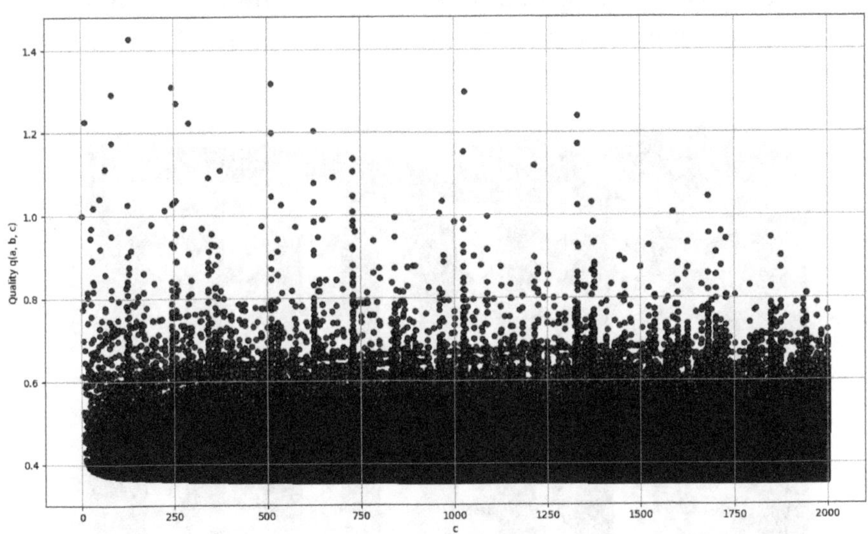

Fig. 6.10   abc conjecture graph of the upper bound of $1 + k$ vs. c for all c's up to 2,000

One of the main messages of this chapter is that connecting whole number analysis with some pictorial representation can lead to important mathematical insights. But it works the other way around as well. Some mathematicians are artists in their own right, and some artists are mathematicians. One of my favorite math artists is Margaret Kepner. Born in South Dakota, and growing up in Japan,

Kepner has a rich background in math, physics and art. Following a career that started as a researcher in radio astronomy, continued as a math and physics high-school teacher, then a software designer for the Federal Bank Reserve, and, until her retirement, in information systems at the International Monetary Fund, she now devotes her time to connecting math and art. Her works have been featured in many formidable venues and won many prizes. My favorite artwork of hers is a Diophantine print called *Prime Goose Chase* (Fig. 6.11). It is a well-thought-out masterpiece of math art. The whole numbers from 1 to 256 are represented by triangles with black or white rectangular backgrounds in 8 columns and 32 rows. The top triangle in the first column on the left depicts the number 1, the triangle

Fig. 6.11   *Prime Goose Chase*, Margaret Kepner, mekvisysuals.net

underneath the number 2, and so on down the column till 32, continuing at the top of the next column with 33, and so on until 256 — the number in the bottom row of the last column. Kepner uses subdivision of the triangles, their rectangular background, and the colour and degrees of shadings of both triangles and rectangles to give much information about the numbers in a visually appealing way.

For example, take the colour of the rectangle backgrounds. For the prime numbers, the colour is black; for the square numbers, white; and for the composite numbers, shades of grey, where the more prime factors the number has, the darker the shade. This is calculated accurately so that the shade for, say, a number with four prime factors, like 210 ($2\times3\times5\times7$), is twice as dark as that for a number with two prime factors, like 6 (2×3). The numbers themselves are colour-coded in the following way. Each prime number is given a different colour — so, 2 is red, 3 is yellow, 5 is green, and so on. Kepner cleverly uses 55 different colours for the primes between 2 and 256 (1 is not prime and is uniquely coloured black) — you'll get to ponder how she does this in the challenges section — so all the triangles with a black background are uniquely uni-coloured.

All the other triangles, the composite numbers — those with black or grey backgrounds — are divided vertically into smaller, equal-area triangles, each indicating a prime factor of the number. For example, the triangle for the number 6 is divided into two triangles, the left one coloured red — the colour of the prime number 2, and the right one yellow — the colour of the prime number 3. If a prime factor appears more than once, its triangle is subdivided horizontally, with the shading darkening as we get lower. For example, the number 12 ($2\times2\times3$) is divided vertically into a red triangle and a yellow triangle, and the red triangle, indicating the prime factor 2, is subdivided into two parts horizontally because there are two prime factors 2. The lower part is a darker red than the upper. The different shading and the horizontal partitioning using lines are visual means to help distinguish the different prime factors. This is especially useful for large square numbers, like 128 ($2 \times 2 \times 2 \times 2 \times 2 \times 2 \times 2$), which is subdivided horizontally into 7 parts, with the shading getting darker as it gets lower. Without the partitioning, the shading, or both, it would be difficult to distinguish 128 from 256, 64, or perhaps even from the prime number 2. Notice the diagonal grey stripes emanating from every sixth triangle? This illustrates a well-known theorem about prime numbers: every prime number larger than 3 is either one below or one above a multiple of 6. Why are there only 256 numbers in *Prime Goose Chase*? Good question! That's just the artist's choice. Kepner's colour and shading schemes were cleverly devised so that, in principle, the artwork can be extended forever, at least theoretically. In practice, distinguishing between the colours and shading quickly becomes too difficult.

Recap

In this chapter, we crossed several paths in number theory and showed how art might help better understand numbers. Beginning with a Diophantine puzzle, we traced its origins in Ancient Egypt math, the life and times of Archimedes and his cattle problem, and Diophantus himself. Along the way, we encountered Egyptian fractions, Pythagorean triples, Ulam's spiral, Pell's equation, and the open problem known as the abc conjecture. Here are some of the main concepts we came across in this chapter.

- *Diophantine equation* — an equation for which you seek to find whole-number solutions.

- *Arithmetica* — Diophantus' famous series of 13 math books.

- *Fermat's last theorem* — a theorem of Pierre de Fermat that appeared in the margin of a printed edition of *Arithmetica*, stating that there is no solution for the equation $a^n + b^n = c^n$ for

$n > 2$. The theorem was proved in 1995 by British mathematician Andrew Wiles.

- *The Rhind Papyrus* — an ancient Egyptian math text written by an Egyptian scribe named Ahmes and found by British archaeologist Henry Rhind in 1858.

- *False assumption* — a problem-solving technique used to solve an equation by making an initial wild guess of the unknown, which is subsequently improved by analysing the outcome.

- *Egyptian mathematics* — The term for the mathematical system used in Ancient Egypt.

- *Hieroglyphics, hieratics, and demotics* — three ancient Egyptian writing systems.

- *Unit fraction* — a fraction where the numerator is 1.

- *Egyptian fractions* — fractions written as a sum of distinct unit fractions.

- *Open problem* — a math problem that has not yet been solved.

- *Erdős–Straus conjecture* — Any fraction whose numerator is 4 can be written as a sum of three unit fractions at the most.

- *Square number* — a number obtained by multiplying a whole number by itself.

- *Scientific notation* — a way of writing large numbers by rounding them up to the nearest power of ten.

- *Pictorial representation of numbers* — the colouring of digits of large numbers or number sequences so that many digits can be seen at once. This helps when we want to search for interesting patterns.

- *Ulam's spiral* — The pattern created when you write the integers in a spiral and mark only prime ones.

- *Pythagorean triple* — three whole numbers that satisfy the Pythagorean theorem $a^2 + b^2 = c^2$.

- *Pell's equation* — An equation of the form $x^2 - ny^2 = 1$, where x and y are sought unknowns and n is a non-square whole number.

- *Coprimes* — numbers with no common divisor except the number 1.

- *Conjecture* — a theorem that has not yet been proved.

- *Distinct prime factors* — the distinct prime factors of a number are the prime numbers that divide it without a remainder.

- *Radical* — the radical of a given number is the product of its distinct prime factors.

- *The abc conjecture* — only a finite number of whole triples (a, b, c) exist where $c > rad(a \times b \times c)^{1+k}$.

- *Prime Goose Chase* — mathematical artwork by Margaret Kepner depicting the prime factorisation of whole numbers.

- *The first 13,689 digits of Pi* — mathematical artwork by Martin Krzywinski depicting the first digits of Pi spiralling outward to fill a circle.

- *Diophantine quadruple (appears in the next section)* — a group of four numbers with the property that if you take any pair of numbers, multiply them, and add 1, you get a square number.

Challenge Yourself!

(1) A treasure chest contains a mix of gold coins and silver coins. The silver coins weigh 10 grams each, while the gold coins weigh 12 grams each. The total weight of the 35 coins in the chest is 390 grams. How many coins of each type are there in the chest?

(2) Write the following fractions as Egyptian fractions. The number of terms in each solution is written in brackets.

(a) $\frac{3}{10}$ (2)

(b) $\frac{2}{11}$ (2)

(c) $\frac{2}{31}$ (3)

(3) Use Euclid's algorithm to generate three primitive Pythagorean triples.

(4) Euclid's algorithm was used to generate the primitive Pythagorean triple: (65, 72, 97). What were a and b?

(5) A Diophantine quadruple is a group of four numbers with the property that if you take any pair of numbers, multiply them, and add 1, you get a square number. 1, 8 and 120 are three of the four numbers in a Diophantine quadruple. The fourth is less than 10. Find it!

(6) Margaret Kepner cleverly uses 55 different colours for the primes between 2 and 256. She does this by starting out with eight bold colours: red, orange, yellow, green, blue, indigo, violet, and purple, assigning each in turn to the first eight primes. The next eight primes get exactly the same colours in the same order, but each colour is now mixed with a small percentage of white to make it slightly paler and distinguishable from its bolder partner, eight numbers before. By doing this, there are now new colours for eight more numbers. This scheme is continued, adding more percentages of white to generate eight paler colours for the next numbers, and so on. For aesthetic reasons, the percentage of white added in each round becomes smaller and smaller. For example, the first eight colours are mixed with 0% white, the next eight with 12.5%, the next eight with 23.4%, the next eight 33.0%, the next eight 41.4%, and so

on. Every round, the percentage of white to be added is 12.5% plus 0.875% of the previous amount added. Hence:

- $23.4 = 12.5 + 0.875 \times 12.5$
- $33.0 = 12.5 + 0.875 \times 23.4$
- $41.4 = 12.5 + 0.875 \times 33.0$

(a) How many rounds are needed to colour the 55 prime numbers between 2 and 256?

(b) What is the percentage of white shade for each of these rounds?

(c) What are the colours of the 11th, 33rd and 48th prime numbers (include the correct white shade percentage)?

(d) How many numbers can be generated using this scheme before the colours all become white? (Assume that you can always distinguish even minute differences in shades.)

(7) Create your own mathematical artwork! You're welcome to send me your creations!

Solutions

(1) If all the coins were silver, the total weight would be $35 \times 10 = 350$ grams. But the total weight is 390 grams, 40 grams more. Since each gold coin adds 2 grams, there must be 20 gold coins and 15 silver coins. Indeed: $15 \times 10 + 20 \times 12 = 390$.

(2) (a) $\frac{3}{10} = \frac{1}{5} + \frac{1}{10}$

    (b) $\frac{2}{11} = \frac{1}{6} + \frac{1}{66}$

    (c) $\frac{2}{31} = \frac{1}{20} + \frac{1}{124} + \frac{1}{155}$

(3) Using Euclids's algorithm, we first choose two coprime integers, a and b, where $a > b$. The first Pythagorean number is $a^2 - b^2$, the second is $2 \times a \times b$ and the third is the square root of the sum

of the squares of the two Pythagorean numbers that we found. Here are my three Pythagorean triples:

- $a = 2$, $b = 1 \rightarrow a^2 - b^2 = 3$; $2 \times a \times b = 4$; $3^2 + 4^2 = 5^2$

  $\rightarrow$ The triple is: **(3, 4, 5)**

- $a = 5$, $b = 1 \rightarrow a^2 - b^2 = 24$; $2 \times a \times b = 10$; $10^2 + 24^2 = 26^2$

  $\rightarrow$ The triple is: **(10, 24, 26)**

- $a = 13$, $b = 11 \rightarrow a^2 - b^2 = 48$; $2 \times a \times b = 286$; $48^2 + 286^2 = 290^2$

  $\rightarrow$ The triple is: **(48, 286, 290)**

(4) This is an interesting problem that can be solved in a few ways. Here's mine. It uses only reasoning and no algebra or guesswork. From the given triple (65, 72, 97), we can infer that:

- $a^2 - b^2 = 65$ or 72

- $2 \times a \times b = 65$ or 72

But since $2 \times a \times b$ is an even number, it can't be 65, so it must be 72, so we now know that:

- $a^2 - b^2 = 65$

- $2 \times a \times b = 72$

If $2 \times a \times b = 72$, then $a \times b = 36$, so (a, b) can be one of the following:

- $(a, b) = (1, 36)$ or $(2, 18)$ or $(3, 12)$ or $(4, 9)$ or $(6, 6)$

We can rule out (2, 18) and (6, 6) straight away because the difference of their squares will be even when it has to be odd (65). (1, 36) can be ruled out because the difference of their squares will certainly be a number larger than 65 — we're subtracting 1 from 36^2, which will be a 3-digit number — and we can rule out (3, 12) because they are not coprime.

(5) The fourth number is 3. Interestingly, there is an infinite number of Diophantine quadruples, yet no Diophantine quintuples (groups of five integers)!

(6) Recall that there are eight colours: red, orange, yellow, green, blue, indigo, violet, and purple.

(a) 55 divided by 8 is 6 with remainder 7, so seven rounds are needed (including the first round that has no white shading).

(b) We'll use the term $S(n)$ to indicate the shade for each round n. We'll also call the first round the zero-th round to make things easier. The shade in the zero-th round, $S(0)$, is 0%. $S(1)$ equals 12.5%. The shade for round n, $S(n) = S(1) + 0.875 \times S(n-1)$. This is known as a recurrence relation, and equations like this can be solved in a few ways. For small numbers, you can do it by hand to get:

- $S(0) = 0\%$

- $S(1) = 12.5\%$

- $S(2) = 23.4\%$

- $S(3) = 33.0\%$

- $S(4) = 41.4\%$

- $S(5) = 48.7\%$

- $S(6) = 55.1\%$

But what would you do if you had to calculate a larger number, like the shading after 18 rounds?

A closer look at the equation gives us an answer. If you write down the recursive equation explicitly for n, you get the sum of a convergent geometric series: $12.5 \times (1 + 0.875 + 0.875^2 + 0.875^3 + \cdots + 0.875^n)$, which, if you remember high-school math, equals: $12.5 \times (1 - 0.875^n)/(1 - 0.875) = 100 \times (1 - 0.875^n)$. Plug in any value of n you like to get the shading of the corresponding round. For $n = 18$, you get: $100 \times (1 - 0.875^{18}) = 90.96\%$.

(c) The remainder after dividing by 8 gives us the place of the colour in the row: red, orange, yellow, green, blue, indigo, violet, and purple (remainder 0 is purple). And the quotient is the n of the shading (when the remainder is 0, the n of the shading should be the quotient -1, because we used the 0 to denote every 8th colour).

- $11 \div 8 = 1$ remainder 3, so the colour is yellow and the shading, $S(1) = 12.5\%$
- $33 \div 8 = 4$ remainder 1, so the colour is red and the shading, $S(4) = 41.4\%$
- $48 \div 8 = 6$ remainder 0, so the colour is purple and the shading, $S(5) = 48.7\%$

(d) An infinite number! The series is a convergent series that reaches 100 at — infinity! You can see this if you compute the answer in Excel. In the first cell, A1 write 12.5. In the second, write the formula $= \$A\$1 + 0.875 * A1$. Then, just drag the formula down the cells. You'll see that the answers get closer and closer to 100 but never reach it (at least — until the computer runs out of place to fit in more decimal digits)!

Bibliography and Further Reading

Browkin, J. (2000). The abc-conjecture. In: *Number Theory. Trends in Mathematics*. Bambah, R. P., Dumir, V. C., and Hans-Gill, R. J. (eds.) Birkhäuser, Basel. https://doi.org/10.1007/978-3-0348-7023-8_5

Gamwell, L. (2016). *Mathematics and Art: A Cultural History*. Princeton University Press.

Hettle, C. (2015). The symbolic and mathematical influence of Diophantus's *Arithmetica. Journal of Humanistic Mathematics*, **5**(1), 139–166.

Imhausen, A. (2016). *Mathematics in Ancient Egypt.* Princeton University Press. https://doi.org/10.1515/9781400874309

Kepner, M. (2013). Prime Goose Chase. In: *Illustrating Mathematics.* David, D. (ed.) American Mathematical Society, pp. 58–59.

Klarreich, E. (2018). Titans of Mathematics clash over epic proof of ABC Conjecture. *Quanta Magazine.* Retrieved August 12, 2024, from https://www.quantamagazine.org/titans-of-mathematics-clash-over-epic-proof-of-abc-conjecture-20180920/.

Krzywinski, M. (2013). 13,689 Digits of Pi. *Martin Krzywinski Digital Art.* Retrieved August 1, 2023, from https://martin-krzywinski.pixels.com/featured/13689-digits-of-pi-martin-krzywinski.html.

O'Connor, J. J. and Robertson, E. F. (1999a). Diophantus of Alexandria. *MacTutor Index.* School of Mathematics and Statistics, University of St Andrews, Scotland. Retrieved August 1, 2023, from https://mathshistory.st-andrews.ac.uk/Biographies/Diophantus/.

O'Connor, J. J. and Robertson, E. F. (1999b). Mathematics in Egyptian Papyri. *MacTutor Index.* School of Mathematics and Statistics, University of St Andrews, Scotland. Retrieved August 1, 2024, from https://mathshistory.st-andrews.ac.uk/HistTopics/Egyptian_papyri/.

Parker, M. (2020). *Humble Pi: When Math Goes Wrong in the Real World.* Penguin Publishing Group.

Vardi, I. (1998). Archimedes' cattle problem. *The American Mathematical Monthly,* **105**(4), 305–319. https://doi.org/10.1080/00029890.1998.12004887

Weisstein, E. W. (2002). Prime spiral. *From MathWorld — A Wolfram Web Resource.* Retrieved August 1, 2023, from https://mathworld.wolfram.com/PrimeSpiral.html.

Wikipedia contributors. (2024). Diophantine equation. *Wikipedia, The Free Encyclopedia.* Retrieved July 29, 2024, from https://en.wikipedia.org/w/index.php?title=Diophantine_equation&oldid=1228869285.

Wikipedia contributors. (2024). Pythagorean triple. *Wikipedia, The Free Encyclopedia*. Retrieved July 30, 2024, from https://en.wikipedia.org/w/index.php?title=Pythagorean_triple&oldid=1234951980.

Wikipedia contributors. (2024, July 25). Pell's equation. *Wikipedia, The Free Encyclopedia*. August 14, 2024, from https://en.wikipedia.org/w/index.php?title=Pell%27s_equation&oldid=1236564612.

Figure Credits

Fig. 1.3 Image of a scytale with the Latin plain text Keiser Augustin hars viktet
[Credit: Luringen, CC BY-SA 3.0,
https://commons.wikimedia.org/w/index.php?curid=1698345]

Fig. 1.9 Portrait of Gerolamo Cardano
[Credit: http://www.storiadimilano.it/Personaggi/Milanesi%20illustri/
cardano_ritratto.jpg, Public Domain,
https://commons.wikimedia.org/w/index.php?curid=64737402]

Fig. 1.16 Route with which to extract the plain text letters from the table in
Fig. 1.15
[Credit: C. M. G. Lee — Own work, CC BY-SA 4.0,
https://commons.wikimedia.org/w/index.php?curid=112059136]

Fig. 2.1 Photograph of Lewis Carroll from 1863
[Credit: Oscar Gustave Rejlander — Unknown source, Public Domain,
https://commons.wikimedia.org/w/index.php?curid=108080]

Fig. 3.4 Carl Friedrich Gauss
[Credit: Christian Albrecht Jensen, — http://archiv.bbaw.de/archiv/
archivbestaende/abteilung-sammlungen/gesamtbestand-des-kunstbesitzes/
gelehrtengemaelde/gelehrtengemalde-seiten/ZIMM-0001.html, Public
Domain, https://commons.wikimedia.org/w/index.php?curid=6886354

Fig. 3.5 Srinivasa Ramanujan
[Credit: Konrad Jacobs, https://opc.mfo.de/detail?photoID=2328, Public
Domain, https://commons.wikimedia.org/w/index.php?curid=111802441]

Fig. 3.6 Jakow Trachtenberg photographed on April 23, 1940, by the
Vienna Gestapo
[Credit: Jakow Trachtenberg — Vienna City and State Archives, Gestapo,
K1 — Gestapo file, CC BY-SA 4.0,
https://commons.wikimedia.org/w/index.php?curid=132474177]

Fig. 3.7 Aaryan Nitin Shukla
[Credit: Kss007in — Own work, CC BY-SA 4.0,
https://commons.wikimedia.org/w/index.php?curid=121070894]

Fig. 4.1 A classic maze
[Credit: Generated using mazegenerator.net, with permission]

Fig. 4.2 Solution to the classic maze
[Credit: Generated using mazegenerator.net, with permission]

Fig. 4.3 Possible restoration of the Labyrinth of Egypt by Italian archaeologist
Luigi Canina
[Credit: Public Domain]

Fig. 4.4 The classic Cretan Labyrinth design
[Credit: JamesJen — Own work, CC BY-SA 3.0,
https://commons.wikimedia.org/w/index.php?curid=7096900]

Fig. 4.5 Edward Burne-Jones painting of Theseus in the Minotaur's Labyrinth
[Credit: Edward Burne-Jones — lgFxdQtUgyzs7Q at Google Cultural Institute,
zoom level maximum, Public Domain,
https://commons.wikimedia.org/w/index.php?curid=29661124]

Fig. 4.7 The Jericho Labyrinth
[Credit: (Humus sapiens) — Own work, Public Domain,
https://commons.wikimedia.org/w/index.php?curid=331193]

Fig. 4.11 The Greek key
[Credit: Beautiful Buildings Pics — Own work, CC BY-SA 4.0,
https://commons.wikimedia.org/w/index.php?curid=83943006]

Fig. 4.12 The Theseus mosaic
[Credit: Carole Raddato from FRANKFURT, Germany — Theseus Mosaic,
discovered in the floor of a Roman villa at the Loigerfelder near Salzburg
(Austria) in 1815, 4th century AD, Kunsthistorisches Museum Vienna, Austria,
CC BY-SA 2.0, https://commons.wikimedia.org/w/index.php?curid=45895870]

Fig. 5.2 The 536 solutions to Archimedes Ostomachion puzzle
    [Credit: Tetsunori Nakayama (tetunori),
    http://openprocessing.org/sketch/1041477,
    https://creativecommons.org/licenses/by-sa/3.0]

Fig. 5.3 Four random solutions to Archimedes' Ostomachion puzzle
    [Credit: Tetsunori Nakayama (tetunori),
    http://openprocessing.org/sketch/1048183,
    https://creativecommons.org/licenses/by-sa/3.0]

Fig. 5.4 Portrait of a Scholar, Domenico Fetti, 1620
    [Credit: Domenico Fetti — http://archimedes2.mpiwg-berlin.mpg.de/
    archimedes_templates/popup.htm, Public Domain,
    https://commons.wikimedia.org/w/index.php?curid=146592]

Fig. 5.5 An Egyptian working on the field using an irrigation pump —
    Archimedes' screw
    [Credit: Zdravko Pečar — Museum of African Art (Belgrade), CC BY-SA 4.0,
    https://commons.wikimedia.org/w/index.php?curid=99668583]

Fig. 5.6 A regular 96-gon is inscribed in a square circumscribed by a
    regular 96-gon
    [Credit: adapted by Yossi Elran from Tomruen,
    https://upload.wikimedia.org/wikipedia/commons/8/82/Regular_
    polygon_96.svg, public domain]

Fig. 5.7 A tangram
    [Credit: Nevit Dilmen — Own work, CC BY-SA 3.0,
    https://commons.wikimedia.org/w/index.php?curid=1798693]

Fig. 5.8 A page from Archimedes Palimpsest where you can see the text of the
    prayer book and the faint underlying original text
    [Credit: The Walters Museum — http://www.archimedespalimpsest.net.,
    CC BY 3.0, https://commons.wikimedia.org/w/index.php?curid=5711744]

Fig. 5.10 Ostomachion puzzles
    [Credit: Meisenstrasse, CC BY-SA 4.0, https://commons.wikimedia.org/wiki/
    File:OstomachionFigures.jpg]

Fig. 5.14 Kate Jones' tricolour Stomachion Puzzle
    [Credit: Joe Marasco, Kadon Enterprises]

Fig. 5.20 The tricolour Ostomachion catalogue
    [Credit: Joe Marasco, Kadon Enterprises]

Fig. 5.23 A proof of the Pythagorean Theorem
    [Credit: JohnBlackburne — Own work, CC BY-SA 3.0,
    https://commons.wikimedia.org/w/index.php?curid=11978281]

Fig. 5.25 Gilroy Song's Pickagram solids shown as related to their tangram
    counterparts
    [Credit: Gil Song, Pickagram]

Fig. 5.26 A Pickagram duck and ship
    [Credit: Gil Song, Pickagram]

Fig. 5.27 The relationship between Pickagram and the Golden ratio
    [Credit: Gil Song, Pickagram]

Fig. 6.1 Page 61 from Diaphantus' Arithmetica
    [Credit: Public Domain,
    https://commons.wikimedia.org/w/index.php?curid=533640]

Fig. 6.2 A fragment of the Rhind Papyrus from the British Museum, London,
    England
    [Credit: unknown (c. 2000 B.C) — Ahmes (scribe),
    http://www.archaeowiki.org/Image:Rhind_Mathematical_Papyrus.jpg
    (https://www.britishmuseum.org/, British Museum), Public Domain,
    https://commons.wikimedia.org/w/index.php?curid=6943889]

Fig. 6.3 Kurt Sethe's compilation of Ancient Egyptian numerals
    [Credit: Florian Cajori — https://en.wikisource.org/wiki/Page:A_History_Of_
    Mathematical_Notations_Vol_I_(1928).djvu/32, Public Domain,
    https://commons.wikimedia.org/w/index.php?curid=84253534]

Fig. 6.4 Pictorial solution of Archimedes' second part of the cattle problem
    [Credit: Yossi Elran]

Fig. 6.5 The first 13,689 digits of pi
    [Credit: Martin Krzywinski]

Fig. 6.6 Ulam's spiral
    [Credit: Morn — Own work, CC0,
    https://commons.wikimedia.org/w/index.php?curid=130815496]

Fig. 6.11 Prime Goose Chase, Margaret Kepner, mekvisysuals.net
   [Credit: Margaret Kepner, mekvisysuals.net]

The author did his best to find and credit the sources for all the images used in this book. All the images in the book that are not in the list above are either in the public domain or were created by the author.